Mission Critical Operations Primer

Mission Critical Operations Primer

Steve Mustard, PE, CAP, GICSP

Notice

The information presented in this publication is for the general education of the reader. Because neither the author nor the publisher has any control over the use of the information by the reader, both the author and the publisher disclaim any and all liability of any kind arising out of such use. The reader is expected to exercise sound professional judgment in using any of the information presented in a particular application.

Additionally, neither the author nor the publisher has investigated or considered the effect of any patents on the ability of the reader to use any of the information in a particular application. The reader is responsible for reviewing any possible patents that may affect any particular use of the information presented.

Any references to commercial products in the work are cited as examples only. Neither the author nor the publisher endorses any referenced commercial product. Any trademarks or tradenames referenced in this publication, even without specific indication thereof, belong to the respective owner of the mark or name and are protected by law. Neither the author nor the publisher makes any representation regarding the availability of any referenced commercial product at any time. The manufacturer's instructions on the use of any commercial product must be followed at all times, even if in conflict with the information in this publication.

The opinions expressed in this book are the author's own and do not reflect the view of the International Society of Automation.

Copyright © 2018 International Society of Automation (ISA)
All rights reserved.

Printed in the United States of America.
10 9 8 7 6 5 4 3 2

ISBN: 978-1-945541-71-1

No part of this work may be reproduced, stored in a retrieval system, or transmitted in any form or by any means, electronic, mechanical, photocopying, recording or otherwise, without the prior written permission of the publisher.

ISA
67 T. W. Alexander Drive
P.O. Box 12277
Research Triangle Park, NC 27709

Library of Congress Cataloging-in-Publication Data in process

To all the engineers and technicians who strive to make the world a better place for everyone, and to those who inspire and develop the next generation to take up the challenge. To my father, John Mustard, an engineer who made the world a better place and inspired me.

Contents

About the Author...**xiii**

Acknowledgments..**xv**

Chapter 1 Introduction....................................**1**
About This Book....................................3
Intended Audience..................................3

Chapter 2 Mission Critical Operations Concepts............**5**
Critical Infrastructure..............................5
Availability, Integrity, and Confidentiality............6
IT–OT Convergence.................................7
Interdependencies in Critical Infrastructure...........8
Review Questions..................................10

Chapter 3 Mission Critical Standards......................**13**
Standards and Regulations..........................13
Common Standards Bodies and Their Key Standards...14
United States Code of Federal Regulations............18
Common Regulatory Bodies and Their Key
 Regulations....................................18
Presidential Policy Directive 21 and the 16 DHS
 Critical Infrastructure Sectors....................21
Review Questions..................................23

vii

Chapter 4 Mission Critical Technology 25
    Introduction 25
    Integrating with Business Systems 26
    Custom and Commercial Off-the-Shelf. 27
    Communications Principles 27
        Open Systems Interconnection (OSI) Model. 27
        IP Addressing 29
        Network Topologies. 32
        LANs, WANs, MANs, and VPNs. 33
        Wired and Wireless 35
        Protocols 37
    Process Control Networks. 44
        Sensors and Actuators. 44
        Control Systems 44
        Control Methodologies 49
        Safety Systems 52
        Historians 52
        Communications Networks in Process Control. . . . 53
        Specialist Communications Networks 55
    Emerging Concepts 56
    Mission Critical Cybersecurity. 58
        Defense in Depth 59
        Cybersecurity Management Systems 61
        Redundancy 63
        Dataflow Security, Integrity, and Reliability 64
        Remote Access. 68
        Cyber Hygiene 69
        Access Control 70
        Malware Prevention and Patching. 70
        System Hardening 70
        Removable Media Control. 71
        Backup and Recovery 72
        Detection and Prevention Systems. 72
    Review Questions. 73

Chapter 5 Operations 81
    Introduction 81
    The Standard Operating Procedure. 82
        SOP Version Control. 82
        The Elements of an SOP 84

	Examples of SOP Presentation 85
	Separation of Duties........................... 85
	Troubleshooting, Repair, and Restoration.............. 87
	Performance Objectives......................... 87
	Troubleshooting Principles 88
	Critical Repairs 90
	Monitoring, Alerting, and Response................. 91
	Logging and Monitoring........................ 91
	Events, Alarms, Alerts, and Notifications.......... 92
	Alarm Priorities 92
	Alarm Handling and Escalation 93
	Event–Alarm Correlation 95
	False Positives and False Negatives.............. 95
	Audit and Maintenance............................. 96
	Configuration Management 97
	Upgrade Management.......................... 98
	Periodic Testing 98
	Auditing...................................... 99
	Change Management............................... 100
	The Change Process........................... 101
	Implementing the Change...................... 102
	Verifying and Finalizing the Change 103
	Life-Cycle Management............................. 104
	Forecasting and Provisioning................... 104
	Commissioning................................ 105
	Support....................................... 106
	Obsolescence Planning 108
	Decommissioning and Disposal 108
	Resource Management 109
	Review Questions................................... 110

Chapter 6 Safety and Physical Security 115

Occupational Safety................................ 115
  Personal Protective Equipment 115
  Lockout/Tagout Procedures.................... 118
  Safety Orientation and Hazardous Operations ... 119
  Job Safety Analysis............................ 120
  Safety Data Sheets 121
  Hazardous Materials Identification System 124
Process Safety...................................... 126

　　　　　　　Hazardous Process Controls. 126
　　　　　　　Fail-Safe Mechanisms 127
　　　　　　　Process Hazard Analysis. 127
　　　　　　　Safety Instrumented Functions and Safety
　　　　　　　　　Integrity Levels. 130
　　　　　　　Hazardous Area Classification. 131
　　　　　Environmental Safety 135
　　　　　　　Emissions and Discharges 135
　　　　　　　Loss of Containment 137
　　　　　　　Safe Handling and Disposal of Materials. 138
　　　　　　　Public Safety 138
　　　　　Physical Security. 139
　　　　　　　Access Control 140
　　　　　　　Intrusion Detection and Prevention 140
　　　　　　　Incident Response 140
　　　　　Review Questions. 141

Chapter 7　Fundamentals of Risk Management 145
　　　　　Definition of Risk 145
　　　　　The Risk Management Cycle 146
　　　　　Risk Management Components 147
　　　　　Quantitative and Qualitative Risk Analysis. 152
　　　　　Process Hazard Analysis. 152
　　　　　Identification of Mission Critical Assets 153
　　　　　Hazard and Threat Assessment. 154
　　　　　Vulnerability Assessment 155
　　　　　Risk Assessment. 156
　　　　　Risk Management Plans 156
　　　　　Review Questions. 158

**Chapter 8　Continuity of Operations and Emergency
　　　　　Response 161**
　　　　　Business Resilience Planning 161
　　　　　　　The Purpose of Business Resilience Planning 161
　　　　　　　Identifying and Prioritizing Essential
　　　　　　　　　Functions and Resources. 164
　　　　　Incident Response 165
　　　　　　　Implementing an IR Plan 165
　　　　　　　Incident Command System. 166
　　　　　　　The Tiered Support Structure. 167

    Regulatory Compliance........................ 168
    Stakeholder Communications 168
Disaster Recovery.................................. 169
    Implementing a DR Plan...................... 169
    Recovery Objectives........................... 169
    Disaster Recovery Drills 171
    Root Cause Analysis and After Action Reviews .. 171
Emergency Management........................... 173
    Phases of Emergency Management............. 173
Review Questions.................................. 175

Appendix A: Answers to Review Questions................. 177
    Chapter 2: Mission Critical Operations Concepts 177
    Chapter 3: Mission Critical Standards................ 178
    Chapter 4: Mission Critical Technology 179
    Chapter 5: Operations 181
    Chapter 6: Safety and Physical Security 183
    Chapter 7: Fundamentals of Risk Management....... 184
    Chapter 8: Continuity of Operations and
        Emergency Response........................... 184

Bibliography **187**
    Technology 187
    Operations.. 188
    Safety and Physical Security....................... 189
    Risk Management................................. 190
    Emergency Response.............................. 190

Index **191**

About the Author

Steve Mustard is an independent automation consultant and a subject matter expert for the International Society of Automation (ISA) and its umbrella association, the Automation Federation. He also is an ISA Executive Board member.

Backed by 30 years of engineering experience, Mustard specializes in the development and management of real-time embedded equipment and automation systems. He serves as president of National Automation, Inc.

Mustard is a recognized authority on industrial cybersecurity, having developed and delivered cybersecurity management systems, procedures, training, and guidance to many critical infrastructure organizations. He serves as the chair of the Automation Federation's Cybersecurity Committee.

Mustard is a licensed Professional Engineer, UK registered Chartered Engineer, a European registered Eur Ing, an ISA Certified Automation Professional® (CAP®), and a certified Global Industrial Cybersecurity Professional (GICSP). He also is a fellow at the Institution of Engineering and Technology (IET), a Senior Member of ISA, a member of the Safety and Security Committee of the Water Environment Federation (WEF), and a member of the American Water Works Association (AWWA).

Acknowledgments

I would like to express my gratitude to the many people who were involved in the long evolution of this book.

To Sarah Carroll and my fellow subject matter experts, for the many enjoyable workshops that helped define the topic of this book.

To Tina Ward, for the original idea for this book and continued support throughout the process.

To Barry Liner and Lisa McFadden, for getting the ball rolling and plotting out the path.

To Leo Staples, Chris Simpson, Ross Coppage, and David Gallagher for their invaluable review comments, based on their extensive knowledge and experience.

To Liegh Elrod, Susan Colwell, and the International Society of Automation (ISA) publications team for all their hard work, turning my material into a professional product.

I am forever indebted to Michael Marlowe and Steve Huffman, without whom none of this would have been possible.

1
Introduction

The term *mission critical* applies to any activity, system, or equipment whose failure can result in the disruption of an organization's operations. Depending on the organization involved, the consequences of failure can be very wide-ranging.

At one extreme, the failure of an online vendor's website can result in a loss of sales. While this could be disastrous to the business concerned, the impact is limited in scope and recovery may not be difficult. Most of us have experienced problems accessing Amazon, Facebook, or Twitter, for instance. While these outages may make the news and can result in a significant financial impact for the business concerned, operations are normally returned promptly and there are few, if any, lasting consequences.

At the other extreme, the failure of control systems in a petrochemical operation could result in injury and loss of life to personnel and the public, as well as harm to the environment from which recovery may be extremely time-consuming, expensive, and difficult. Consider, for example, the Deepwater Horizon accident in 2010. Eleven people lost their lives on the day of the accident, which resulted in the largest oil spill ever in U.S. waters.

The negative impact to the environment is still being experienced in the Gulf region of the United States and will be for many years to come. As of 2017, the costs arising from the accident, including financial settlements and fines, exceeds $62 billion.

Mission critical operations can be impacted by a wide variety of factors: hardware or software failures, network communications problems, accidental damage or disruption, or natural disasters. One factor making the news regularly is cyber attack. High-profile incidents have impacted household names such as Sony, Target, eBay, P.F. Chang's, and Domino's Pizza. In these cases, confidential information was stolen, resulting in the need for major disaster-recovery activities. On Christmas Day 2014, a group known as the Lizard Squad successfully brought down the Xbox Live and PlayStation networks. As a result, 48 million Xbox Live subscribers and 110 million PlayStation users were unable to access their respective networks, causing major disruptions on one of the year's biggest days of demand.

In the industrial space, reports indicate a 10-fold increase in the number of successful cyber attacks on infrastructure control systems since 2000. This is partly a consequence of advances in control systems, enabling them to be integrated into the business environment. Although this has proven to be a huge benefit for businesses, allowing better visibility of process information in near real-time, the increased connectivity has exposed new vulnerabilities that can be targeted by attackers. The connection between industrial (or operational technology—OT) and information technology (IT) systems has created problems for both types of systems. For instance, in Germany in December 2014, a steel mill was attacked and the blast furnace suffered major damage. The origin of the attack was the business network, where the attackers were able to navigate to the control system network and disrupt the emergency shutdown systems that were designed to prevent major damage to the plant.

There are many potential cyber attackers, such as hackers seeking to prove their capabilities, criminals seeking access to financial gain, and state-funded operations designed to damage another state's activities. As a result, mission critical systems must be designed and operated to cope with accidental and deliberate incidents. In addition, the management of such systems requires an enhanced level of diligence, as the nature and source of threats is always changing.

A whole culture of mission critical operations specialists has emerged. These specialists understand the threats, risks, and consequences of failure. Although they may focus on areas such as robust IT network design, control system security, control room operations, and alarm handling, these specialists will normally have a broad understanding of *all* key aspects of mission critical systems. No other career requires so many different aspects to be brought together in one role.

About This Book

This book is a primer on mission critical operations. The objective of the book is to provide a high-level overview of key concepts. There are many aspects to mission critical operations and each one can be studied in further depth. It is not the author's intent to repeat the detail already provided in other books, a list is provided for further reading.

Intended Audience

This book is intended for those who need a high-level understanding of the key concepts in mission critical operations, including students on entry-level programs and those beginning their careers.

2

Mission Critical Operations Concepts

Critical Infrastructure

Modern society is dependent on the underlying critical infrastructure that provides power, water, waste disposal, transportation, financial services, and emergency services. Mission critical operations are essential to the smooth and continued availability of these services.

Technology plays a major role in modern mission critical operations management. There are two distinct forms of technology that exist in mission critical organizations:

- **Information technology (IT)** – This includes computing equipment and systems, networking equipment and systems, and associated processes required to manage a typical business. Most mission critical organizations will have an IT function or department that is responsible for the technology and processes.

- **Operational (or operations) technology (OT)** – This includes the systems, devices, and associated processes that are required to manage physical processes and plants, such as control valves, engines, conveyors, and other machines. In general, OT is the responsibility of an engineering function or department.

Availability, Integrity, and Confidentiality

Management of IT and OT involves many similar aspects, but there are some crucial differences. One fundamental difference is the relative importance of the following factors, as shown in Figure 2-1:

- **Availability** – Making sure the system or information is there when it is needed.
- **Integrity** – Making sure the system is operating correctly or that information is complete and not corrupted.
- **Confidentiality** – Protecting information from falling into the wrong hands.

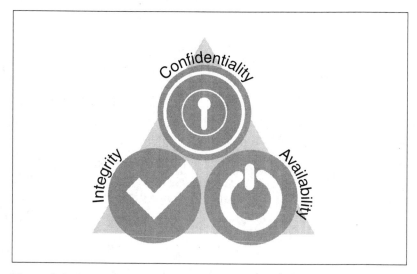

Figure 2-1. Availability, Integrity, and Confidentiality

All three aspects are important in OT and IT systems, but the relative importance of each varies depending on the type of system.

OT is responsible for monitoring and controlling industrial processes; and failure could have a significant impact on safety, production, and the environment. For OT, the relative order of importance is:

- Availability
- Integrity
- Confidentiality

In other words, OT places priority on the continuous and accurate operations of systems rather than securing confidential information. However, a breach of confidentiality can indirectly lead to loss of availability and integrity. In addition, access to OT systems could provide access to proprietary information, such as details of a novel manufacturing process or product recipe or contents.

For IT, the relative order of importance is:

- Confidentiality
- Integrity
- Availability

In other words, IT places a priority on securing data and protecting it from being corrupted.

IT–OT Convergence

IT and OT have traditionally operated as distinctly separate entities, with very little direct interaction. However, business demands have been changing, for instance:

- The need for greater visibility of plant operation in real time.
- The need to identify productivity improvements through analysis of plant data.
- The need to optimize operations by integrating with other organizations in the supply chain.
- The need to increase standardization to reduce costs.

These changing business demands have resulted in the need to achieve closer integration of the two entities. Technology advances, such as reducing the cost of processing power and the availability of lower cost communications options, have made this closer integration possible.

This closer integration has yielded significant benefits for organizations; however, it has also increased risks, especially in the area of cybersecurity:

- The interconnected nature of operations now makes many more systems and devices vulnerable to deliberate attack or accidental misuse.
- The use of common equipment in IT and OT environments increases the potential impact of a cybersecurity incident, as attackers exploit the same vulnerabilities in both environments.

A key part of the role of mission critical operations professionals is to be able to bridge the gap between the IT and OT worlds, by understanding the drivers and the technology involved in both.

Interdependencies in Critical Infrastructure

Mission critical organizations in the critical infrastructure do not exist in isolation. There is a heavy interdependency of organizations. Figure 2-2 shows a simplified example, illustrating

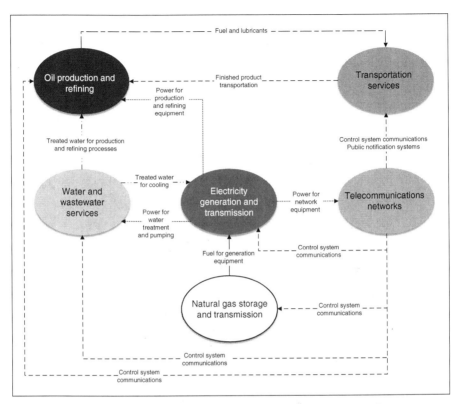

Figure 2-2. The Interdependency of Critical Infrastructure Operations

some interdependencies between various critical infrastructure organizations.

Electricity plays a key role in all critical infrastructure organizations, but electricity (i.e., its generation and transmission) is itself dependent on other organizations—such as telecommunications providers, natural gas providers, and water and wastewater companies—to operate.

Even within a mission critical organization, interdependencies are at play. For example, an organization's heating, ventilation, and air conditioning (HVAC) system is an essential dependency for the reliable operation of an IT data center.

For mission critical organizations to operate continuously, systems and procedures need to take into account potential failures, both within the organization and outside. Such organizations also need to consider dependencies that other organizations have on them.

Key considerations are:

- **Resilience** – The organization must anticipate potential failure events and provide protection against them. For instance, it is likely that backup power generation facilities are deployed to ensure continuous operation in the event of a failure of the main supply.

- **Reliability** – The organization must use components that are designed for continuous operation in the specific operating environment.

- **Redundancy** – The organization must ensure that backup facilities and capabilities are in place. For example, in order to ensure that a failure of the Internet service provider connection between sites does not result in a loss of service, the organization should deploy a separate backup service—for instance, using cellular.

- **Response** – The organization must have plans and procedures in place to provide the ability to respond to events and limit effects. The plans and procedures must include the capability to restore services in a timely manner (e.g., by creating regular backups).

Review Questions

2.1 What does the term *availability* mean in relation to systems?

    A. The system is operating correctly or that information is complete and not corrupted.

B. The system is protecting information from falling into the wrong hands.

C. The system or information is there when it is needed.

D. The system provides accurate or dependable information.

2.2   Which of the three aspects are most important for IT and OT systems?

A. For IT systems it is confidentiality; for OT systems it is availability.

B. For IT systems it is availability; for OT systems it is confidentiality.

C. For IT systems it is integrity; for OT systems it is availability.

D. For IT systems it is availability; for OT systems it is integrity.

2.3   What is one reason for the convergence of IT and OT systems?

A. To use the same technology throughout a facility

B. To adopt higher-reliability IT components in OT systems

C. To simplify and reduce operations and maintenance tasks

D. The need for greater visibility of plant operation

2.4   What does the term *redundancy* mean in relation to systems?

A. The system or information is there when it is needed.

B. The system continues to operate in the event of a component failure.

C. The system is operating correctly or the information is complete and not corrupted.

D. The system provides accurate or dependable information.

2.5 What does the term *reliability* mean in relation to systems?

A. The system is operating correctly or the information is complete and not corrupted.

B. The system continues to operate in the event of a component failure.

C. The system operates continuously without interruption.

D. The system or information is there when it is needed.

3
Mission Critical Standards

Standards and Regulations

In mission critical organizations, there will be a need to understand and comply with various standards and regulations. This will vary depending on the industry sector and, in some cases, the organization's customer base (e.g., a power company providing power to a military base may have different standards and regulations than one providing power to domestic users).

Standards differ from regulations. A standard is established by consensus, and approved by a recognized body. Subject matter experts from the relevant sector of industry, as well as from government, come together to develop standards. A standards development organization (SDO) manages the process. A country's national standards body (e.g., American National Standards Institute—ANSI—in the United States) approves SDOs.

Standards are developed to help improve compatibility by agreeing on specific requirements. Compliance with standards is voluntary, but usually highly desirable. For example, defining the frequencies used by wireless devices enables all vendors to manufacture products that are interoperable. However, some contractual agreements may mandate compliance with particular standards. Standards are sometimes referred to as best practice or recommended practice.

Regulations are similar to standards in that they define specific requirements. Key differences between regulations and standards are:

- Regulations carry the force of law.

- Regulations normally prescribe what is required, whereas standards identify how to achieve that requirement.

A governing authority prescribes regulations. Unlike standards, regulation compliance is mandatory. Examples of operations in which regulations are enforced include electricity generation and distribution, chemical production and storage, and water treatment and distribution.

There are many standard and regulatory bodies, and many standards and regulations produced and maintained by them. The following sections provide some of the most commonly referenced standards and regulations and their relevant bodies (Tables 3-1 and 3-2).

Common Standards Bodies and Their Key Standards

Table 3-1. Common Standards Bodies and Key Standards

Body	Overview of Body	Key Standards
American Society of Heating, Refrigerating and Air-Conditioning	A professional member and standards organization focused on	• Standard 34 – *Designation and Safety Classification of Refrigerants* • Standard 55 – *Thermal Environmental Conditions for Human Occupancy*

Table 3-1. Common Standards Bodies and Key Standards (*Continued*)

Body	Overview of Body	Key Standards
Engineers (ASHRAE)	building systems, energy efficiency, indoor air quality, refrigeration, and sustainability technologies.	• Standards 62.1/62.2 – Ventilation standards • Standard 90.1 – *Energy Standard for Buildings Except Low-Rise Residential Buildings* • Standard 135 – *BACnet – A Data Communication Protocol for Building Automation and Control Networks* • Standard 189.1 – *Standard for the Design of High Performance, Green Buildings Except Low-Rise Residential Buildings*
American Society of Mechanical Engineers (ASME)	A professional member and standards organization focused on mechanical engineering.	• *ASME Boiler and Pressure Vessel Code* (BPVC) • A17 Series – Elevators and escalators • B31 Series – Piping and pipelines • B16 Series – Valves flanges, fittings, and gaskets
American Petroleum Institute (API)	A trade organization and standards organization dedicated to the petroleum industry.	• API RP 520 – *Sizing, Selection, and Installation of Pressure-Relieving Devices in Refineries* • API RP 521 – Guide for Pressure Relief and Depressuring Systems • API RP 14 – *Recommended Practices for Systems for Offshore Production Platforms*
International Electrotechnical Commission (IEC)	An international standards organization that prepares and publishes international standards for all electrical, electronic, and related technologies.	• IEC 61511 – *Functional Safety – Safety Instrumented Systems for the Process Industry Sector* • IEC 61508 – *Functional Safety of Electrical/Electronic/Programmable Electronic Safety-Related Systems* • IEC 60529 – Ingress protection marking (see also NEMA) • IEC 62443 – *Security for Industrial Automation and Control Systems*

Table 3-1. Common Standards Bodies and Key Standards (*Continued*)

Body	Overview of Body	Key Standards
Institute of Electrical and Electronics Engineers (IEEE)	A professional member and standards organization focused on electrical and electronic engineering.	• IEEE 802.3 – IEEE Standard for Ethernet • IEEE 802.11 – Wireless networking – "WiFi" • IEEE 802.15.2 – *Bluetooth* • IEEE 802.15.4 – *Wireless Sensor/Control Networks – "ZigBee"*
Internet Engineering Task Force (IETF)	An international nonprofit standards organization that develops and promotes voluntary Internet standards.	• RFC 791 – *Internet Protocol* • RFC 793 – *Transmission Control Protocol* • RFC 854 – *Telnet Protocol Specification* • RFC 959 – *File Transfer Protocol (FTP)* • RFC 1157 – *A Simple Network Management Protocol (SNMP)* • RFC 2460 – *Internet Protocol, Version 6 (IPv6) Specification* • RFC 4291 – *IP Version 6 Addressing Architecture*
International Society of Automation (ISA)	A professional member and standards organization focused on automation engineering.	• ISA-5 – Symbols, diagrams, and identification • ISA-18 – Alarm systems and instrument signals • ISA-88 – *Batch Control* • ISA-95 – *Enterprise-Control System Integration* • ISA99 – ISA-62443 *Security for Industrial Automation and Control Systems* series (see also IEC 62443 for international version)
Interagency Security Committee (ISC)	A collaborative organization led by Department of Homeland Security that provides security leadership to the nonmilitary federal community.	• November 2016/2nd Edition – *The Risk Management Process: An Interagency Security Committee Standard* • February 2013/1st Edition – *Items Prohibited from Federal Facilities: An Interagency Security Committee Standard* • December 2015/1st Edition – *Best Practices for Planning and Managing Physical Security Resources: An Interagency Security Committee Guide*

Table 3-1. Common Standards Bodies and Key Standards (*Continued*)

Body	Overview of Body	Key Standards
		• February 2015/1st Edition – *Presidential Policy Directive 21 Implementation: An Interagency Security Committee White Paper*
International Organization for Standardization (ISO)	An international standards organization that prepares and publishes international standards related to a wide range of subjects.	• ISO 9000 series – Quality management systems • ISO 14000 series – Environmental management systems • ISO 27000 series – Information technology – Security techniques – Information security management systems • ISO 15408 – *Common Criteria for Information Technology Security Evaluation*
National Electrical Manufacturers Association (NEMA)	A trade association of electrical equipment and medical imaging manufacturers in the United States.	• ANSI 60529 – Ingress protection marking (see also IEC 60529) • NEMA 250-2003 – *Enclosures for Electrical Equipment 1,000 Volts Maximum*
National Fire Protection Association (NFPA)	A trade association that creates and maintains private, copyrighted standards and codes for usage and adoption by local governments.	• NFPA 70 – *National Electrical Code* (NEC) • NFPA 79 – *Electrical Standard for Industrial Machinery*
National Institute of Standards and Technology (NIST)	A standards laboratory and a nonregulatory agency of the U.S. Department of Commerce.	• NIST Cybersecurity Framework • NIST 800 series – Computer security standards • Federal Information Processing Standard (FIPS) 140-2 – *Security Requirements for Cryptographic Modules*

United States Code of Federal Regulations

The U.S. Code of Federal Regulations (CFR) defines a list of rules and regulations published in the Federal Register by the executive departments and agencies of the federal government of the United States. The CFR is divided into 50 titles that represent broad areas subject to federal regulation, such as:

- Title 10 – *Energy*
- Title 18 – *Conservation of Power and Water Resources*
- Title 29 – *Labor*
- Title 49 – *Pipeline and Hazardous Materials Safety Administration, Department of Transportation*

Regulations are commonly referred to by their title (e.g., Title 10, Title 10 CFR, or 10 CFR). Where applicable, a section or subsection may be included in the reference (e.g., 29 CFR 1910.119).

The Office of the Federal Register and the Government Publishing Office publish an annual edition of the CFR. In addition to this annual edition, the CFR is published in an unofficial format online on the Electronic CFR website.

Common Regulatory Bodies and Their Key Regulations

Table 3-2. Common Regulatory Bodies and Key Regulations

Body	Overview of Body	Key Standards
Occupational Safety and Health Administration (OSHA)	Agency of the U.S. federal government that regulates working conditions.	• 29 CFR 1910 – *Occupational Safety and Health Standards* • 29 CFR 1910 Subpart I – *Personal Protective Equipment* • 29 CFR 1910.119 – *Process Safety Management of Highly Hazardous Chemicals* (process safety management—PSM)

Table 3-2. Common Regulatory Bodies and Key Regulations (*Continued*)

Body	Overview of Body	Key Standards
Environmental Protection Agency (EPA)	Agency of the U.S. federal government that protects human health and the environment.	• PL 84-159 – *Air Pollution Control Act* • PL 88-206 – *Clean Air Act* • PL 93-523 – *Safe Drinking Water Act* • PL 95-217 – *Clean Water Act* • PL 91-190 – *National Environmental Policy Act* • PL 94-469 – *Toxic Substances Control Act* • PL 97-425 – *Nuclear Waste Repository Act* • Section 112(r) 1990 *Clean Air Act* amendments – *Risk Management Plan (RMP) Rule*
Federal Energy Regulatory Commission (FERC)	Agency of the U.S. federal government that regulates the transmission and wholesale sale of electricity and natural gas in interstate commerce, and regulates the transportation of oil by pipeline in interstate commerce.	Title 18 CFR – *Conservation of Power and Water Resources*
Nuclear Regulatory Commission (NRC)	An independent agency of the U.S. government tasked with protecting public health and safety related to nuclear energy.	Title 10 CFR – *Energy*
North American Electric Reliability Corporation (NERC)	A not-for-profit international regulatory authority that assures the reliability and security of the bulk power system in North America.	Critical Infrastructure Protection (CIP): • CIP-002-5.1a – *Cyber Security—BES Cyber System Categorization* • CIP-003-6 – *Cyber Security – Security Management Controls* • CIP-004-6 – *Cyber Security – Personnel & Training* • CIP-005-5 – *Cyber Security – Electronic Security Perimeter(s)* • CIP-006-6 – *Cyber Security – Physical Security of BES Cyber Systems*

Table 3-2. Common Regulatory Bodies and Key Regulations (*Continued*)

Body	Overview of Body	Key Standards
		• CIP-007-6 – *Cyber Security – System Security Management* • CIP-008-5 – *Cyber Security – Incident Reporting and Response Planning* • CIP-009-6 – *Cyber Security – Recovery Plans for BES Cyber Systems* • CIP-010-2 – *Cyber Security – Configuration Change Management and Vulnerability Assessments* • CIP-011-2 – *Cyber Security – Information Protection Related Information* • CIP-014-2 – *Physical Security*
Pipeline and Hazardous Materials Safety Administration (PHMSA)	A U.S. Department of Transportation (DOT) agency responsible for developing and enforcing regulations for the safe, reliable, and environmentally sound operation of U.S. pipeline transportation.	• 49 CFR 100–185 – Pipeline and Hazardous Materials Regulations
Food and Drug Administration (FDA)	A U.S. federal agency, part of the Department of Health and Human Services, responsible for protecting and promoting public health through the control of food, tobacco products, medications, biopharmaceuticals, blood transfusions, medical devices, electromagnetic radiation-emitting devices, cosmetics, animal feed, and veterinary products.	• 21 CFR 11 – *Electronic Records and Electronic Signatures* (ERES)

Table 3-2. Common Regulatory Bodies and Key Regulations (*Continued*)

Body	Overview of Body	Key Standards
Securities and Exchange Commission (SEC)	An independent agency of the U.S. federal government, it holds primary responsibility for enforcing the federal securities laws.	• Pub.L. 107–204 – *Sarbanes–Oxley Act* (SOX)
Payment Card Industry (PCI) Security Standards Council	Comprised of members of Visa, MasterCard, American Express, Discover, and JCB, it manages the *Payment Card Industry Data Security Standard*.	• *Payment Card Industry Data Security Standard* (PCI DSS)
Department of Health and Human Services (HHS)	A cabinet-level department of the U.S. federal government responsible for protecting the health of all Americans and providing essential human services.	• 45 CFR 160 – *Health Insurance Portability and Accountability Act of 1996* (HIPAA)
Department of Homeland Security (DHS)	A cabinet department of the U.S. federal government responsible for anti-terrorism, border security, immigration and customs, cybersecurity, and disaster prevention and management.	• 6 CFR 27 – *Chemical Facility Anti-Terrorism Standards* (CFATS) • 44 USC chapter 35, subchapter III – *Federal Information Security Modernization Act* (FISMA)

Presidential Policy Directive 21 and the 16 DHS Critical Infrastructure Sectors

Presidential Policy Directive (PPD-21) is titled, "Critical Infrastructure Security and Resilience." It was issued 12 February 2013 and states:

> It is the policy of the United States to strengthen the security and resilience of its critical infrastructure against both physical and cyber threats. The Federal Government shall work with critical infrastructure owners and operators and [state, local, tribal, and territorial] SLTT entities to take proactive steps to manage

risk and strengthen the security and resilience of the Nation's critical infrastructure, considering all hazards that could have a debilitating impact on national security, economic stability, public health and safety, or any combination thereof.

PPD-21 supersedes Homeland Security Presidential Directive 7 (HSPD-7). PPD-21 calls for a federal government approach to secure critical infrastructure that includes three "strategic imperatives":

- Refine and clarify functional relationships across the federal government to advance the national unity of

Table 3-3. Critical Infrastructure Sectors and Their Assigned Government Agencies

Critical Infrastructure Sector	Sector-Specific Agency (SSA)
Chemical Sector	Department of Homeland Security (DHS)
Commercial Facilities Sector	DHS
Communications Sector	DHS
Critical Manufacturing Sector	DHS
Dams Sector	DHS
Defense Industrial Base Sector	Department of Defense
Emergency Services Sector	DHS
Energy Sector	Department of Energy
Financial Services Sector	Department of the Treasury
Food and Agriculture Sector	Co-SSAs: Department of Agriculture and Department of Health and Human Services (HHS)
Government Facilities Sector	Co-SSAs: DHS and General Services Administration
Healthcare and Public Health Sector	HHS
Information Technology Sector	DHS
Nuclear Reactors, Materials, and Waste Sector	DHS
Transportation Systems Sector	Co-SSAs: DHS and Department of Transportation
Water and Wastewater Systems Sector	Environmental Protection Agency

effort to strengthen critical infrastructure security and resilience.

- Enable effective information exchange by identifying baseline data and systems requirements for the federal government.

- Implement an integration and analysis function to inform planning and operations decisions regarding critical infrastructure.

Under PPD-21, 16 critical infrastructure sectors are defined and specific government agencies are assigned responsibility for them, as shown in Table 3-3.

Review Questions

3.1 What is the key difference between a standard and a regulation?

    A. Regulations are produced by the government but standards are produced by industry.

    B. Regulations are updated annually, whereas standards are updated more frequently.

    C. Compliance with regulations is mandatory but compliance with standards is optional.

    D. Regulations must be purchased but standards are free to use.

3.2 Which standards body is responsible for IEC 62443, *Security for Industrial Automation and Control Systems*?

    A. Department of Homeland Security (DHS)

    B. International Society of Automation (ISA)

C. National Institute of Standards and Technology (NIST)

D. Interagency Security Committee (ISC)

3.3 Which standards body is responsible for the Simple Network Management Protocol (SNMP) standard?

A. Institute of Electrical and Electronic Engineers (IEEE)

B. International Standards Organization (ISO)

C. International Electrotechnical Commission (IEC)

D. Internet Engineering Task Force (IETF)

3.4 Which regulatory authority is responsible for the Critical Infrastructure Protection (CIP) regulations?

A. Pipeline and Hazardous Materials Safety Administration (PHMSA)

B. North American Electric Reliability Corporation (NERC)

C. Department of Homeland Security (DHS)

D. Nuclear Regulatory Commission (NRC)

3.5 Which sector-specific agency is responsible for the Water and Wastewater Systems Sector?

A. Department of Homeland Security (DHS)

B. Environmental Protection Agency (EPA)

C. Department of Agriculture (USDA)

D. General Services Administration (GSA)

4
Mission Critical Technology

Introduction

In general, mission critical systems control and monitor machinery or other equipment, keep the underlying physical processes in their normal operating envelope, and ensure that hazardous conditions are managed safely and reliably. Although the type of equipment, machinery, or physical process can vary considerably (e.g., wastewater pumps, packing machines, assembly robots, and oil valves), the elements of mission critical systems can be generalized as follows:

- Sensors to measure process conditions, such as temperature or pressure, or machine states, such as valve position or robot orientation.
- Actuators to adjust process values, such as electric motors or valves.
- Controllers (e.g., programmable logic controller—PLC).
- Data loggers (e.g., remote terminal unit—RTU) to interface to the sensors and actuators.

- Specialist analyzer systems that monitor particular process parameters. Examples include gas analyzers that measure the quality of a product for fiscal or quality-control purposes, and condition monitoring to monitor vibration on rotating equipment, such as turbines.

- Central control systems to allow operators to monitor and remotely control equipment and processes.

- Communications networks to allow the controllers and data loggers to communicate with each other and central control systems.

Integrating with Business Systems

Mission critical systems were designed for the specific function of controlling and monitoring machinery or other equipment. As such the systems were used in stand-alone, isolated situations in remote facilities, such as oil and gas platforms or factories.

As businesses have continued to drive efficiency and productivity improvements, there has been a trend to integrate mission critical systems with other business systems. For example:

- Integrating with asset management and work management systems allows for planning of routine maintenance within the organization, based on equipment operating hours or other conditions.

- Integrating equipment vendors into the condition-monitoring process allows for the vendor to perform a more comprehensive analysis and repair service.

- Integrating with materials procurement systems allows for just-in-time delivery of raw materials based on real-time measurements.

Custom and Commercial Off-the-Shelf

The technology behind the components in mission critical systems has traditionally been highly specialized and customized. However, cost and efficiency demands, coupled with the drive to integrate equipment with business systems, have resulted in the use of commercial off-the-shelf (COTS) hardware and software, such as Windows operating systems and standard PC-based processing cards.

Communications Principles

The communications networks that are most commonly used in mission critical systems are based on the same principles as those used in information technology (IT) networks.

Open Systems Interconnection (OSI) Model

The Open Systems Interconnection Reference Model is a series of standards published in 1984 by the International Organization for Standardization (ISO) and the International Telegraph and Telephone Consultative Committee (CCITT), now known as the Telecommunications Standardization Sector of the International Telecommunications Union (ITU-T). The standards are usually referred to as the *Open Systems Interconnection Model*, the *OSI Reference Model*, or the *OSI model*, and they consist of the Basic Reference Model, or seven-layer model, and a set of specific protocols. The OSI model defines a standard for the architecture of networking systems, such as those used in business networks, process control networks, and the Internet.

The OSI model divides network communications into seven layers:

- Physical
- Data Link

- Network
- Transport
- Session
- Presentation
- Application

Data is transmitted to and from devices through these layers by means of protocols and services specific to each layer. Figure 4-1 shows how data flows through the model.

A modified version of this model, called the Transmission Control Protocol/Internet Protocol (TCP/IP) model, contains four layers. The Application layer of the TCP/IP model is a consolidation of the Application, Presentation, and Session layers of the OSI model. The Transport layer in the TCP/IP model is unchanged from the OSI model. The Network layer of the OSI

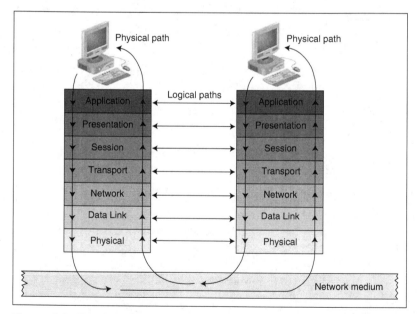

Figure 4-1. The OSI Model

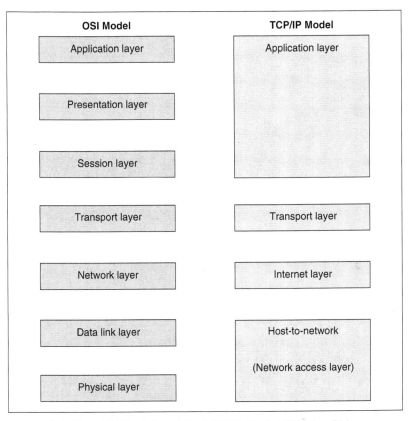

Figure 4-2. The OSI Model and the TCP/IP Model Side-by-Side

model is renamed as the Internet layer in the TCP/IP model. The Network Access or Interface layer of the TCP/IP model is a consolidation of the Data Link and Physical layers of the OSI model. Figure 4-2 shows the relationship between the two models.

IP Addressing

An IP address is an address that uniquely identifies a device on an IP network. There are two standards for IP addressing: IPv4 and IPv6. Figure 4-3 shows the structure of an IPv4 address. In IPv4, the address is made up of 32 binary digits, or bits, which can be divisible into a network portion and host portion. The 32 bits are broken into four octets (1 octet = 8 bits). The value in each octet ranges from 0 to 255 decimal, or 00000000 to 11111111

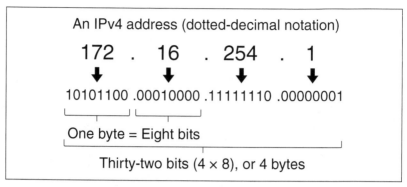

Figure 4-3. Basic Structure of an IP Address

binary. Each octet is converted to decimal and separated by a period (dot), for example, 172.16.254.1.

Using IPv4, it is possible to create 4,294,967,296 (2^{32}) possible unique addresses. Some of these addresses are reserved (e.g., addresses in the range 192.168.x.x, 172.16.x.x, and 10.x.x.x are called *nonroutable addresses* and are reserved for use on internal networks).

In the early days of the Internet, the network and host portions of the address format were created to allow for a more fine-grained network design. The first three bits of the most significant octet of an IP address were defined as the *class* of the address. Three classes (A, B, and C) were defined for addressing. In class A, 24 bits of host addressing allows for 16,777,216 (2^{24}) unique addresses. In class B, only 16 bits of host addressing are available, reducing the number of unique addresses to 65,536 (2^{16}). In class C, only 256 (2^8) unique addresses are possible due to there being only 8 bits for the host address. The three classes are shown in Figure 4-4.

The class approach to network addressing did not prove to be scalable as the Internet grew; and in 1993 the classless interdomain routing (CIDR) method was introduced to replace it.

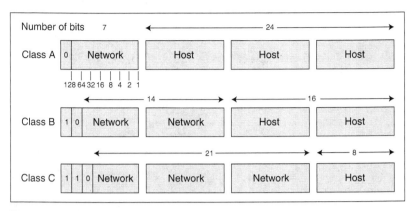

Figure 4-4. Classes of IP Addresses

In IPv4, the CIDR notation is written as the first address of a given network followed by the bit-length of the network portion of the address. For example, 192.168.1.0/24 means that there is an address range that starts at 192.168.1.0 and has 256 unique addresses up to 192.168.1.255 (the /24 signifies that the network portion of the address is 24 bits, leaving 8 bits for the host address, which yields 2^8 or 256 addresses).

In 2015, all 16,777,216 IPv4 addresses were allocated leaving none available for future use. Anticipating this issue, plans to replace the standard were developed. IPv6 is the current address standard for the Internet, although IPv4 addressing continues to be supported in parallel and will be for the foreseeable future. All new network equipment must support IPv6. Figure 4-5 shows the IPv6 address format.

IPv6 uses a 128-bit address, allowing 2^{128}, or approximately 3.4×10^{38}, addresses. The CIDR notation for the IPv6 address is similar to that of IPv4 addresses. For example, the IPv6 address 2001:db8::/32 denotes an address block starting at 2001:0db8:0000:0000:0000:0000:0000:0000 with 2^{96} addresses (having a 32-bit routing prefix denoted by /32, leaving 96-bits for host addresses).

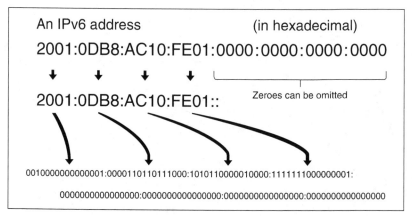

Figure 4-5. The IPv6 Address Format

Network Topologies

The hierarchical internetworking model is a network design that divides a larger network into three parts: *core*, *distribution*, and *access*. Each part provides different services. Figure 4-6 shows the relationship between the three parts.

The access layer is where hosts connect to the network. The aggregation, or distribution, layer is where the access layer is connected; it provides connectivity to adjacent access layer switches and data center rows, and, in turn, to the top of the tree, known as the *core*. The core layer provides routing services to other parts of the network, as well as to services outside of the network. A variation of this design, called the *collapsed core*,

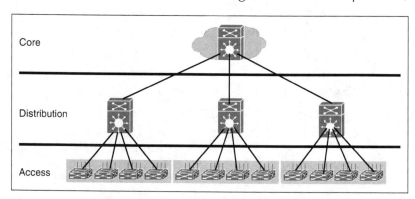

Figure 4-6. The Hierarchical Internetworking Model

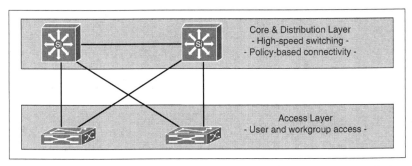

Figure 4-7. The Collapsed Core Model Combines Parts of the Hierarchical Model

combines two parts into one where this is more suitable. This is shown in Figure 4-7.

In leaf and spine architectures, the three parts become the *core*, *spine*, and *leaf*; and they provide greater connectivity between the layers to afford increased reliability. The leaf and spine architecture is shown in Figure 4-8.

LANs, WANs, MANs, and VPNs

Networks form a key part of mission critical systems. There are various levels of networks that mission critical systems will use and interact with; these are shown in Figure 4-9:

- A local area network (LAN) is a group of connected devices, typically within one building.

- A metropolitan area network (MAN) is a larger network that usually spans several buildings in the same city or

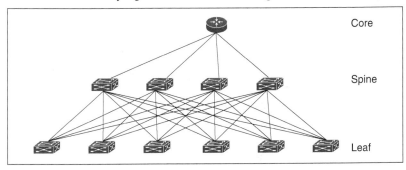

Figure 4-8. The Leaf and Spine Model

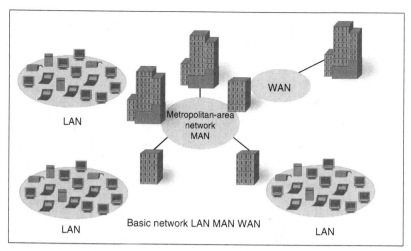

Figure 4-9. LANs, MANs, and WANs

town. Examples of MANs can be found on college campuses and business campuses.

- A wide area network (WAN) connects several LANs and is not restricted to a geographical location.

One key network security control is the virtual private network (VPN), shown in Figure 4-10. When configured correctly, routers and firewalls ensure that it is not possible to access devices on a private network from outside that network. When access to these devices is required, a VPN is a recognized safe method of doing so. A VPN enables users to send and receive data across shared or public networks as if their devices were directly connected to the private network. As a result, they benefit from the functionality, security, and management policies of the private network.

A VPN provides remote access in a secure way, by enforcing:

- **Authentication** – To ensure that only authorized users can access the network

- **Encryption** – To ensure that all data transferred over the VPN cannot be intercepted by unauthorized users

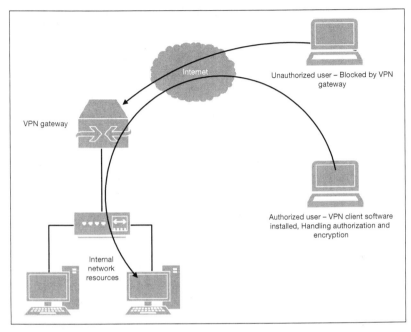

Figure 4-10. Virtual Private Network (VPN)

A VPN gateway is a connection point that facilitates the remote connections to the private network. This device provides the authentication and encryption functionality necessary to provide secure remote access. The VPN gateway may be a hardware device or software installed on a remote machine.

Wired and Wireless

Wi-Fi is now widely used in control systems. While it provides increased flexibility over wired connections, encryption and authentication must be carefully managed to avoid introducing security risks.

The Institute of Electrical and Electronics Engineers (IEEE) defines the standards for Wi-Fi access, including encryption. Wi-Fi encryption began with Wired Equivalent Privacy (WEP), which was embodied in the original IEEE 802.11 standard. Wi-Fi Protected Access (WPA), embodied in IEEE 802.11i, has since superseded this protocol. WEP should not be used, as it is insecure due to the short encryption key, only 40 bits, which allows hackers to break the encryption relatively quickly. WPA uses 128-bit keys and WPA-2 uses 152-bit keys, both of which are so time-consuming to break that they can be considered secure.

The IEEE 802.1X standard defines authentication for Wi-Fi networks. In 802.1X, authentication involves three parties: a *supplicant*, an *authenticator*, and an *authentication server*. This is shown in Figure 4-11.

The supplicant is a client device, such as a laptop, that is connecting to the network. The authenticator is a network device, such as a wireless access point. The authentication server runs software supporting the Remote Authentication Dial-In User Service (RADIUS) and Extensible Authentication Protocols (EAP).

The supplicant provides credentials, such as user name/password or digital certificate, to the authenticator, and the

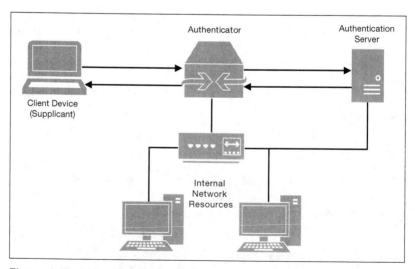

Figure 4-11. 802.1X Wi-Fi Authentication

authenticator forwards the credentials to the authentication server for verification. The supplicant is not allowed access through the authenticator to the protected side of the network until the supplicant's identity has been validated and authorized by the authentication server.

Protocols

Protocols define rules and conventions for communication between network devices. Protocols include mechanisms for devices to identify and make connections with other devices, as well as defining rules that specify how data is packaged into messages that are sent and received. Some protocols support message acknowledgment and data compression designed for reliable and high-performance network communication. Hundreds of computer network protocols have been developed, each designed for specific purposes and environments.

There are four key groups of protocols, shown in Figure 4-12:

- Networking protocols
- Routing protocols
- Redundancy protocols
- Application layer protocols

The networking protocol category includes those protocols that are involved with the configuration of the network.

The Address Resolution Protocol (ARP) and the Reverse Address Resolution Protocol (RARP) operate at the Network layer and are used to map physical, or media access control (MAC), addresses to logical, or IP addresses. A network device wishing to obtain a physical address broadcasts an ARP request on the network. The device on the network that

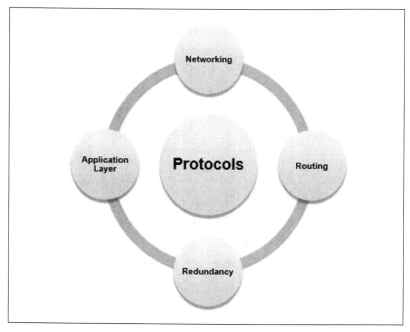

Figure 4-12. Protocols

has the IP address in the request then replies with its physical hardware address. RARP can be used by a device to discover its own IP address. In this case, the device broadcasts its physical address and a RARP server replies with the device's IP address.

DHCP is short for Dynamic Host Configuration Protocol. DHCP is an Application layer protocol for assigning dynamic IP addresses to devices on a network. With *dynamic addressing*, a device can have a different IP address every time it connects to the network. In some systems, the device's IP address can even change while it is still connected. Using dynamic IP addressing enables users to move workstations and printers without having to change their network configuration. The other method of addressing is called *static addressing*. When a device is assigned a static IP address, the address does not change. Static

IP addresses normally matter more when external devices or websites need to remember an IP address. One example is a remote access solution that is configured to trust only certain IP addresses for security purposes.

The domain name system (DNS) is an Internet service that translates domain names into IP addresses. The Internet is really based on IP addresses, but it is much easier to remember names than IP addresses. Every time you use a domain name, a DNS must translate the name into the corresponding IP address. For example, the domain name www.example.com might translate to 198.105.232.4. Assignment of IP addresses to domain names is managed by an organization called the Internet Corporation for Assigned Names and Numbers (ICANN) and its registered agents, called *domain registrars*.

The Internet Control Message Protocol (ICMP) is a Network layer protocol that sends error messages indicating, for example, that a requested service is not available or that a device could not be reached. ICMP can also be used to relay query messages. Many commonly used network utilities are based on ICMP messages. The traceroute command is implemented by transmitting IP messages with specially set header fields and by looking for specific messages in response. The ping utility is implemented using the ICMP *echo request* and *echo reply* messages.

A routing protocol specifies how routers communicate with each other, sharing information that enables them to select routes between any two devices on a computer network. Special algorithms determine the specific choice of primary and failover routes. Each router has knowledge only of networks attached to it directly. A routing protocol shares this information first among immediate neighbors, and then throughout the network. In this way, routers gain knowledge of the overall topology of the network.

With the Routing Information Protocol (RIP), routers periodically exchange entire routing tables that keep track of routes to particular network destinations. With larger and larger networks, the sharing of whole tables becomes inefficient and RIP also has a limit of 15 hop counts that can be accommodated, which limits the scalability of this protocol. As a result, RIP is being replaced by other more efficient and expansive protocols.

Open Shortest Path First (OSPF) is a link-state routing protocol that calls for the sending of link-state advertisements (LSAs) to all other routers within the same hierarchical area. Information on attached interfaces, metrics used, and other variables are included in OSPF LSAs. As OSPF routers accumulate link-state information, they use the Shortest Path First (SPF) algorithm to calculate the shortest path to each node.

The Enhanced Interior Gateway Routing Protocol (EIGRP) is Cisco's proprietary routing protocol, based on the Interior Gateway Routing Protocol (IGRP). EIGRP is a distance-vector routing protocol, with optimizations to minimize routing instability incurred after topology changes, and the use of bandwidth and processing power in the router. Most of the routing optimizations are based on the Diffusing Update Algorithm (DUAL), which guarantees loop-free operation and provides fast router convergence.

RIP and OSPF are examples of Interior Gateway Protocols. These are protocols that operate within specific network domains. Exterior Gateway Protocols (EGP) exchange routing information between domains. The Border Gateway Protocol (BGP) is an EGP. BGP is a robust and scalable routing protocol, which is why it is used by Internet service providers to share information about their networks. It uses a multitude of parameters, known as *attributes*, to define routing policies and maintain stable communications.

Redundancy protocols are used to make very high availability networks. The Spanning Tree Protocol (STP) is an old redundancy protocol that allows a network design to include redundant links to provide automatic backup paths if an active link fails, without the need for manual intervention. STP exchanges Bridge Protocol Data Unit (BPDU) packets with other switches on the network to detect loops, and shuts down interfaces where BPDUs have returned via a loop in the network. Default STP recovery times can be slow (30–50 seconds); however, newer versions, such as Rapid STP, have reduced this to milliseconds.

The Media Redundancy Protocol (MRP) is a newer redundancy protocol that allows groups of Ethernet switches to overcome any single failure, with recovery time much faster than achievable with the Spanning Tree Protocol. MRP uses a ring topology detection mechanism with heartbeat messages sent out from a ring master to determine a link failure, and is commonly used in industrial Ethernet applications.

MRP also operates at the Physical layer and is a direct evolution of the HiPER-Ring protocol used by Hirschmann switches since 2003. MRP is supported by several commercial industrial Ethernet switches. MRP is also adopted as an Internet standard.

The Virtual Router Redundancy Protocol (VRRP) is designed to avoid an individual router on a network being a single point of failure. VRRP dynamically assigns routing responsibility to one of several configured VRRP routers on a LAN, called the *master*. If the master fails, one of the backup routers or switches becomes the new master within a few seconds. This is done with minimum VRRP traffic and without any interaction with the hosts.

Each Application layer protocol has its own port number. Network devices use this port number, along with an IP address,

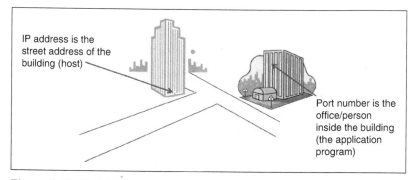

Figure 4-13. IP Addresses and Port Numbers Are Like Street Addresses and Office Numbers

to identify the service to be used on a particular device. An IP address is similar to the mailing address for a building. Once at that particular building, the port number can be thought of as the number or person's name associated with a particular office. Internet standards define standard port numbers for well-known network services that are used throughout the Internet. Custom or nonstandard services can have their own port, although this would only be known within a specific user network. This is shown in Figure 4-13.

There are many Application layer protocols. These are some of the most widely used:

- **File Transfer Protocol (FTP)** – This protocol is used to send and receive files. In an FTP exchange, there is a server and a client. The client connects to the server and can send files to the server or receive files from the server using different commands within the protocol.

- **Secure Shell (SSH)** – This protocol provides emulation of a traditional computer terminal. It is used to execute commands on remote machines. It includes security measures that protect the users against anyone with malicious intent.

- **Telnet** – This protocol is similar to SSH; however, unlike SSH, it offers no security measures because Telnet was designed to work within a private network and not across a public network. One of the major differences is that all the data is transmitted in plain text, including passwords. SSH uses encryption to make it harder for other attackers to obtain confidential information, such as a user password and other relevant information. Telnet also omits a safety measure called *authentication* that is included in SSH. Authentication ensures that the source of the data is still the same device and not another computer. Without authentication, another person can intercept the communication and do what they wish.

- **Simple Mail Transfer Protocol (SMTP)** – This is a protocol for sending email messages between servers. Most email systems that send mail over the Internet use SMTP to send messages from one server to another. The messages can then be retrieved with an email client using other standard protocols.

- **Hypertext Transfer Protocol (HTTP)** – This is the underlying protocol used by the World Wide Web. HTTP defines how messages are formatted and transmitted, and what actions web servers and browsers should take in response to various commands. For example, when you enter a URL in your browser, this actually sends an HTTP command to the web server, directing it to fetch and transmit the requested web page.

- **Network Time Protocol (NTP)** – This assures accurate synchronization to the millisecond of computer clock times in a network of computers. NTP synchronizes client workstation clocks to the U.S. Naval Observatory Master Clocks in Washington, D.C., and Colorado Springs, Colorado. NTP sends periodic time requests to servers, obtaining server time stamps and using them to adjust the device's clock.

Industrial protocols that operate on networks have their own port numbers. For example, Modbus uses port 502 by default, although it is possible to change this port in many Modbus applications.

Process Control Networks

Sensors and Actuators

Process control networks exist to facilitate communication between control systems and the distributed sensors and actuators that interact with these systems.

Traditionally, sensors and actuators were connected to control systems using analog or digital signals. For example, a sensor measuring temperature, pressure, and flow would provide a corresponding current or voltage proportional to the measurement. An example is shown in Figure 4-14.

For an actuator, the control system would provide a current or a voltage representing the desired setting for the control. With digital signals, the sensor would indicate a door open, a valve closed, or other similar state by a 1 or 0 and the control system would send a 1 or 0 to control a relay or similar device. Digital signals can also be used to represent cumulative values of flow or other similar readings by using pulses to represent quantity. For example, a water meter might generate 1 pulse to indicate 1 gallon of water has flowed through it.

Control Systems

The term *control system* covers a wide variety of solutions that have evolved over many years to meet specific requirements.

Figure 4-14. A Typical Flow Meter, Providing an Analog Current Representation of the Measured Value

The distributed control system (DCS) is widely used in process control facilities, such as oil and gas platforms, refineries, and food and beverage plants. A typical DCS is shown in Figure 4-15.

A DCS typically includes several controllers that communicate with field devices, and a server that displays plant status in a series of formats:

- **Plant overview** – Sometimes referred to as a mimic, these are graphical representations of the plant status, showing, for example, tank levels and valve status in a visual form. An example is shown in Figure 4-16.

- **Trend display** – Time-series data such as flow, temperature, or pressure are shown in graphical form for a given time, including for the last 2 hours, 24 hours, or 1 week.

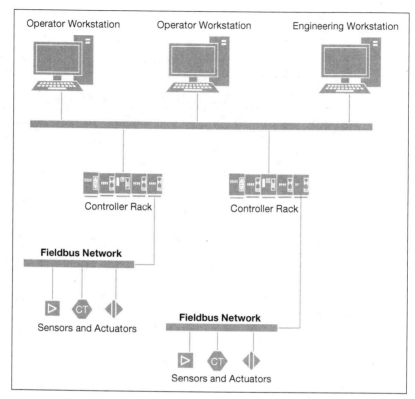

Figure 4-15. A Typical Distributed Control System (DCS)

- **Alarm list** – A list of alarms, such as high temperature and low pressure, are displayed in a tabular form to allow the operations personnel to make decisions about issues.

The information is displayed via one or more human-machine interfaces (HMIs). These provide the primary view into a mission critical system's operation. HMIs vary from vendor to vendor and from industry sector to industry sector. A typical control room showing HMIs is in Figure 4-17.

HMIs tend to be located in control rooms, away from the main process. For local monitoring and control, operator

Chapter 4 – Mission Critical Technology   47

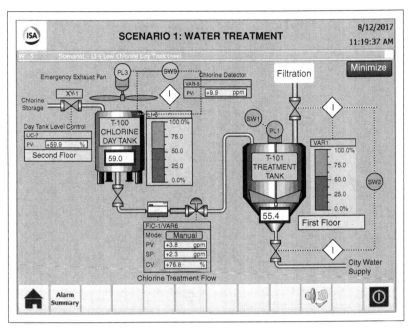

Figure 4-16. An Example of a Plant Mimic Provided by a DCS

interface terminals (OITs) are deployed at key points. An OIT may show similar information to an HMI (e.g., mimics, trends, and alarms) but it will be specific to the area it is monitoring. For instance, pump equipment may have an OIT to allow the operator to set operational values and display status, and this pump information may also be visible on the main HMIs in the control room. Figure 4-18 shows an operator using an OIT.

A supervisory control and data acquisition (SCADA) system is virtually identical, functionally, to the DCS. The main difference between a SCADA system and a DCS is that a SCADA system typically communicates with field devices over a wide geographic area, whereas a DCS will normally communicate with devices on a local network. Another major difference is that SCADA system field devices usually communicate directly with servers, rather than controllers. These controllers tend to be high availability, redundant devices,

Figure 4-17. An Operator Monitoring a Process via HMIs

meaning that a DCS is generally better suited for high availability requirements.

DCS and SCADA systems communicate with field devices called remote terminal units (RTUs) or programmable logic

Figure 4-18. An Operator Checks the Settings of a Process Using an OIT

controllers (PLCs). The functionality of these two field devices is similar, in that they are connected to analog and digital sensors and actuators, and they communicate with DCS and SCADA systems. The major differences between a PLC and an RTU are:

- A PLC is generally designed to communicate over local networks, whereas an RTU is designed to communicate over wide area networks. For instance, a PLC might communicate with a DCS in a refinery, while an RTU might communicate with the same DCS from a remote well miles away.

- A PLC has a programmable capability allowing it to run local programs that read sensors and control actuators. An RTU typically stores the data it collects from sensors and sends it back to the control system.

As technology has developed, the boundary between a PLC and an RTU is less clear. Many modern RTUs now support PLC-like programming and many PLCs now provide for remote communications and data logging. A typical device is shown in Figure 4-19.

Control Methodologies

A major function of control systems is to control industrial processes, such as:

- Water treatment processes
- Oil or gas well operations
- Robotic manufacturing or packaging processes

The methods to control these and other processes vary considerably, but several basic principles apply throughout. In

Figure 4-19. A Typical Modern Device Combining RTU and PLC Functionality

general, the purpose of control is to achieve and maintain a particular setting (or outcome) in a process or subprocess. For example, it may be necessary to maintain the fluid in a tank at a particular level, or temperature, or both. This is achieved using sensors to determine the current level (or temperature) and driving actuators (such as pumps or heaters) until the required settings, called *set points*, are achieved.

Open-loop and *closed-loop control* are two distinct control methodologies that can be applied to various processes. In open-loop control, the control system drives actuators to achieve particular set points but does not adjust the actuators based on sensor readings. This simple form of control is easy to implement but is not desirable in most industrial processes, where some degree of accuracy is necessary for reliable, repeatable, and safe operation. Figure 4-20 shows open- and closed-loop control block diagrams.

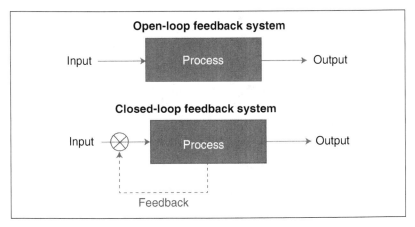

Figure 4-20. Open- and Closed-Loop Control

In contrast, closed-loop control continuously monitors sensor readings and uses these, in conjunction with set points, to maintain process outputs at, or close to, the desired set points. There are many closed-loop control algorithms that exist, and their use varies depending on the type of process; the most common is the proportional-integral-derivative (PID) algorithm.

In the PID algorithm, the proportional element adjusts the input to the process in proportion to the difference between the process output and the set point (this difference is called the *error*). The integral element adjusts the same input in proportion to the integral of the error (i.e., how much the error has changed over time). The derivative element adjusts the input in proportion to the derivative of the error (i.e., how fast the error is changing).

Engineers determine the values of the proportional, integral, and derivate elements based on a combination of modeling, calculation, and experimentation. In some cases, the integral or derivative elements, or even both, are excluded (in which case the control algorithm is called *PD*, *PI*, or *proportional*, respectively).

Safety Systems

A safety instrumented system (SIS), also known as an emergency shutdown (ESD) system or safety shutdown system (SSDS), is present in hazardous plants, such as oil and gas or chemical processing. The SIS monitors the entire control system network, including the DCS or SCADA system and is responsible for shutting down the plant in the event of a potential emergency, such as dangerous pressures or temperatures being detected, or failure of the primary control system.

The current trend is to combine DCS and SIS functionality into one integrated control and safety system (ICSS), although some users still prefer to keep independent DCS and SIS functions.

Historians

A historian is a database designed to store the type of data that process control systems collect. Whereas conventional databases store relational data (e.g., customer contact details and the customer's associated orders), historians are designed to store time-series data, (e.g., flow and pressure readings collected every minute). Example time-series data is shown in Figure 4-21.

Specialist historian databases are designed to efficiently store and retrieve the large volume of real-time data that is collected in a typical process control system.

The DCS or SCADA system is used to monitor the process control system in real time, whereas the historian is typically used to analyze process data to identify longer-term trends. For example, vibration data collected over long periods of time may be analyzed to identify equipment failures before they occur.

SITE	TAG	TIMEBASE	TIMESTAMP	VALUE	ENG UNITS
M161808	P1	1M	10/11/16 18:01	151.2	PSI
M161808	P1	1M	10/11/16 18:02	151.4	PSI
M161808	P1	1M	10/11/16 18:03	152.1	PSI
M161808	P1	1M	10/11/16 18:04	151.5	PSI
M161808	P1	1M	10/11/16 18:05	151.1	PSI

Figure 4-21. An Example of Time-Series Data

Communications Networks in Process Control

Unlike other branches of technology, mission critical technology requires very high-performance processing equipment and communications networks.

Mission critical technology is used to control real-world processes, such as gas processing, electricity distribution, and water treatment. Mission critical systems must be capable of responding in real-time to changes in processes to ensure safe and repeatable operations. Communication networks are a key aspect of mission critical systems.

Analog and digital sensors are still widely used in process control networks and show no sign of disappearing. However, it is now common to connect these sensors to devices, which then communicate with control systems via communications networks.

Process control networks have existed in one form or another for almost 50 years. As a result, they utilize many forms of physical communications networks, including serial and Ethernet-based options. Control systems and associated devices support a variety of communications protocols that operate over these networks. Examples of common communications protocols are:

- **Modbus** – This is one of the oldest and most popular communications protocols in process control networks. Modbus was originally created by the process control company Modicon (now part of Schneider) for their own devices but was adopted as an open standard and is implemented by all major process control equipment vendors. Modbus was originally designed to operate on serial connections, such as RS-232 (for point-to-point communications) or RS-485 (for point-to-multipoint communications). A version of the protocol exists for use on Ethernet networks. It is called Modbus/TCP.

- **DNP3** – This is version 3 of the Distributed Network Protocol, a communications protocol used extensively in the electricity and water industries. Originally created by GE Harris, the protocol is more extensive than Modbus and is designed to allow varying levels of functionality to suit different devices from a simple intelligent electronic device (IED) to a fully functional RTU. Recent additions to the protocol include authentication to provide a more secure communications option.

- **EtherNet/IP** – This is one of the newest and most popular communications protocols and is specifically designed around Ethernet networks. Like DNP3, it provides a more extensive range of features than Modbus.

- **PROFINET** – Short for Process Field Net, this is another modern Ethernet-based communications protocol. There are three options available: Standard TCP/IP, Real Time (PROFINET RT), and Isochronous Real Time (PROFINET IRT). The choice depends on how deterministic the communications needs to be. The PROFIBUS & PROFINET International (PI) group maintains the PROFINET standard, along with PROFIBUS (Process Field Bus), a related communications standard that uses proprietary high-speed serial connections.

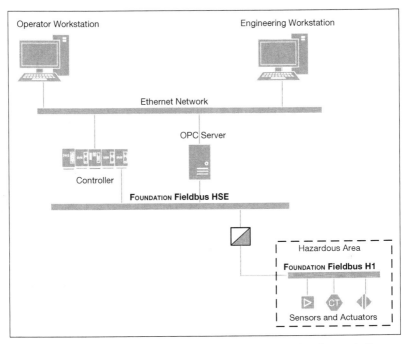

Figure 4-22. An Example of a FOUNDATION Fieldbus Implementation

- FOUNDATION **Fieldbus** – This is a serial communications protocol designed for plant or factory automation. There are now two newer variants of this protocol: FOUNDATION Fieldbus H1, a high-speed version of the basic protocol; and high-speed Ethernet (HSE), which is suitable for use in industrial networks. Figure 4-22 shows an example of a FOUNDATION Fieldbus network.

Specialist Communications Networks

Process control networks may also utilize several specialist physical communications options, sometimes particular to specific industry sectors. Examples of these specialist communications options are:

- **HART** – The Highway Addressable Remote Transducer (HART) protocol operates over analog sensor connections.

It allows users to configure or calibrate an analog sensor using the same connection that is used to provide the analog measurement. The HART digital signals are modulated over the current signal to avoid the need for separate connections.

- **BACnet** – The Building Automation and Control Network protocol (BACnet) was designed to allow communication of building automation and control systems, such as those involved in heating, ventilating, and air-conditioning control (HVAC), lighting control, access control, and fire detection systems.

- **CAN** – The Controller Area Network (CAN) protocol (governing CAN bus architecture) is designed for use in vehicles. It was originally intended to reduce the wiring required in vehicles, allowing multiple devices to communicate over the same wires. The ISO 11898 series of standards, *Road Vehicles – Controller Area Network (CAN)*, are important standards for on-board vehicle diagnostics.

- **AS-i** – The Actuator Sensor Interface (AS-i) is designed to allow the simplified connection of sensors and actuators to PLCs and RTUs. Using sensors and actuators based on AS-i simplifies the wiring of equipment.

Emerging Concepts

Mission critical technology is constantly evolving. Advances in the use of COTS products, integration with business networks, and the development of specialist communications networks and protocols have taken place in recent years. At the time of writing, there are several new emerging concepts in mission critical systems:

- **IIoT** – The Industrial Internet of Things (IIoT) is a term coined to collect all the component parts of modern industrial control and monitoring systems: intelligent sensors and devices and machine-to-machine communications. The major difference between traditional systems and those in the IIoT is that the IIoT is designed to require less user interaction, as decisions and interactions take place at the device level.

- **Virtualization** – This allows a single physical server device to support multiple server environments in parallel. The benefits are significant in terms of hardware cost savings, as well as reduced physical space, cooling, and power. Many mission critical systems, such as DCS and SCADA, are now designed to run in virtualized environments.

- **Cloud computing** – The concept behind cloud computing is to share resources in a centrally accessible location, which may be operated within the organization or by an external third party. The cloud refers to the fact that users do not generally know what is behind the resources (i.e., what type of servers or where they are located) that they are accessing, only that they are available for them to use. Organizations use cloud-computing resources to run applications as well as for storing large volumes of data. DCS and SCADA vendors now offer cloud-based solutions, although many users still believe it is safer and more secure to run such systems within their facilities.

- **Big data and analytics** – Analyzing data is not new; organizations have been looking for trends in data for decades. However, technology enhancements—such as lower cost storage, faster networks, and more intelligent devices—have provided organizations with an opportunity to analyze bigger volumes of data more quickly than ever. With big data and analytics, the objective is to collect as much data as possible and then use complex algorithms and models to identify patterns that would otherwise be unseen. In some mission critical systems, the results of such analysis can be fed back to operational users to allow them to make better real-time decisions.

- **Open source** – The Linux operating system is now widely used in mission critical and other business systems. Linux is an example of open source, a development model that promotes universal access to a product's source code. The concept is that a large and diverse group of developers contribute to the maintenance and enhancement of the product, as opposed to traditional software developed and sold by commercial companies. However, several commercial companies exist to provide professionally repackaged and supported products (such as Red Hat Enterprise Linux) based on open-source components.

Mission Critical Cybersecurity

As with all other technology-based systems, cybersecurity presents a major threat to the safe and reliable operation of mission critical systems. Threats arise from:

- Accidental or unintentional actions of an employee or contractor
- Deliberate actions of a disgruntled employee or contractor
- Actions from external sources, such as terrorists, organized crime, or nation states
- Environmental incidents

Possible cybersecurity incidents include:

- Introduction of malware into control system equipment, causing a loss of control or loss of visibility of the plant
- Unauthorized access, resulting in deliberate or accidental operation of the plant
- Inability to recover systems after equipment failure or an environmental incident

While theft of confidential information from IT systems is a major concern, this is usually not so critical with operational technology (OT) systems. However, the theft of confidential information about the OT system, which could be used to mount an attack, is a concern.

Defense in Depth

The term *defense in depth* is derived from a proven military strategy successfully used for many centuries. In defense in depth, a defender uses multiple protections to provide continued resistance to attackers. In the case of military strategy, this might involve using multiple fortifications or different types of protection. Figure 4-23 shows Dover Castle, a very early example of the defense-in-depth concept.

In safety-related operations the concept of the barrier model is the equivalent of defense in depth. The barrier model identifies the various protections and mitigations that exist in a system or process to reduce the likelihood of an event causing an incident, such as a fire or explosion. The barrier model is also known as the *Swiss cheese model*, because it is recognized that the barriers do not guarantee 100% protection, and so contain holes. As with defense in depth, the barrier model aims to reduce the

Figure 4-23. An Early Example of the Defense-in-Depth Concept—Dover Castle

likelihood of an incident by providing multiple, diverse protections. The barrier model is shown in Figure 4-24.

In cybersecurity the defense-in-depth approach is similar to the military and safety-related versions. It identifies all the protections that are in place to reduce the likelihood of a security

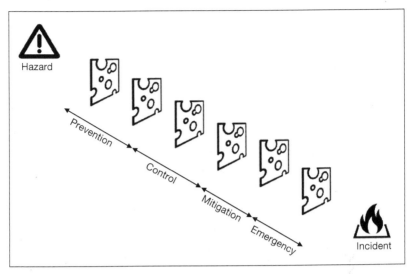

Figure 4-24. The Barrier, or Swiss Cheese, Model

event, such as malware entering a control system and causing an incident. The cybersecurity defense-in-depth approach is shown in Figure 4-25.

Typical defense-in-depth elements include:

- Malware protection and patching
- Access control
- Backup and recovery
- Control of removable media
- Intrusion detection and prevention

Cybersecurity Management Systems

Mission critical organizations must implement a cybersecurity management system if they are to confidently mitigate the risk of cybersecurity incidents. Several methodologies and

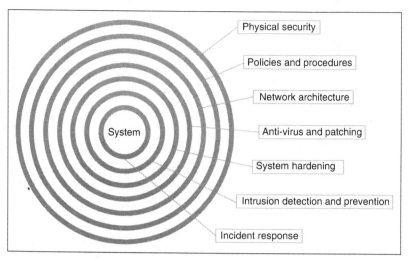

Figure 4-25. Defense in Depth as Applied to Cybersecurity

frameworks exist to support the development of a cybersecurity management system. The National Institute of Standards and Technology (NIST) *Framework for Improving Critical Infrastructure Cybersecurity* (more commonly known as the *NIST Cybersecurity Framework* or NIST CSF) was developed in 2013 in response to President Obama's Executive Order 13636, "Improving Critical Infrastructure Cybersecurity." This framework, shown in Figure 4-26, identifies five high-level activities that mission critical organizations must do:

1. **Identify** – Create an inventory of all cyber assets and understand what these assets do, such as what hardware and software are used, how the equipment is connected, and what external connections exist.

2. **Protect** – Create the defense-in-depth protections, such as access control policies and procedures, malware protection and patching procedures, and backup and recovery strategies.

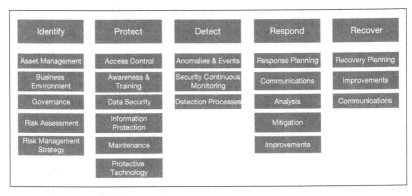

Figure 4-26. NIST Cybersecurity Framework

3. **Detect** – Put in place methods, automatic and manual, to detect unusual activity or unauthorized access.

4. **Respond** – Establish procedures to deal with cybersecurity events.

5. **Recover** – Ensure that procedures are in place before and after cybersecurity events to ensure that it is possible to restore normal operation.

The International Society of Automation (ISA) developed ANSI/ISA 62443, which is the international standard for cybersecurity of industrial automation and control systems. This standard is referenced extensively throughout the *NIST Cybersecurity Framework*. It provides more details on how organizations should go about implementing their cybersecurity management systems.

Redundancy

Failure of hardware or software components can result in a major failure of a mission critical system. Cybersecurity threats, such as malware or targeted denial-of-service attacks, can cause a failure of components, as can inherent vulnerabilities, such as hardware or software flaws.

Where failure of a component cannot be tolerated, redundant design is required. There are several types of redundant design:

- **Cold standby** – Although not a redundant system in the true sense, the availability of spare components that can be rapidly replaced in the event of a failure does provide a basic level of response.

- **Warm standby** – In this scenario, duplicate components are running alongside the live equipment and can be replaced more quickly than in the cold standby scenario. However, there is still some loss of service while the replacement takes place.

- **Hot standby** – This scenario minimizes the downtime experienced during component failure. As with warm standby, duplicate components are running alongside the live equipment. In the hot standby scenario, the duplicate/standby component communicates with its live counterpart and detects failure (or an overall system controller monitors all components and detects failure), at which point the standby component takes over.

The level of complexity in the redundancy design can vary considerably. Safety critical systems, such as aircraft control systems, typically have triplicated components and a "voting" system that makes decisions based on the status of two out of three of the components. Figure 4-27 shows an example of a dual-redundant controller.

Dataflow Security, Integrity, and Reliability

Encryption is a technique for transforming data in such a way that it becomes unreadable—so that, even if someone gains access to the data, they would still need to decrypt it in order to read it. The data to be encrypted, also called *plaintext* or *cleartext*,

Figure 4-27. A Typical Redundant Controller Installation

is transformed using an encryption key, which is a value that is combined with the original data to create the encrypted data, also called *ciphertext*. The same encryption key is used at the receiving end to decrypt the ciphertext and obtain the original cleartext.

In addition to helping ensure that data does not get read by the wrong people, encryption can also ensure that data is not altered in transit, and can verify the identity of the sender.

There are three main types of encryption used in networks:

1. Symmetric (or private-key) encryption
2. Asymmetric (or public-key) encryption
3. Hybrid encryption

In symmetric encryption, the key used to encrypt and decrypt the message must remain secure; therefore, the alternate

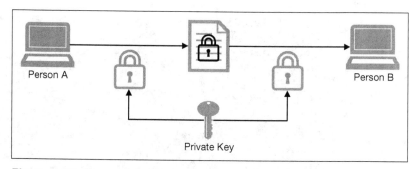

Figure 4-28. Symmetric Encryption

name is *private-key encryption*. Anyone with access to the encryption key can decrypt the data. Using symmetric encryption, a sender encrypts the data with the key, sends the data, and the receiver uses the same key to decrypt the data. This is shown in Figure 4-28.

Examples of symmetric key algorithms are:

- Data Encryption Standard (DES), developed in 1977
- Triple DES (3DES), developed as an improvement to DES
- Advanced Encryption Standard (AES), which supersedes DES and 3DES and is the worldwide standard for data encryption today

Asymmetric encryption is shown in Figure 4-29, and is different from symmetric encryption because it uses two keys: one for encryption and one for decryption. The encryption key is known as the *public key*, as it is freely available to everyone to encrypt messages. As a result, asymmetric encryption is also known as *public-key encryption*.

The decryption key is known as the *private key* because it is not shared. In asymmetric encryption, if person A wants to send a confidential message to person B, person A encrypts the message with person B's public key. The resulting message is

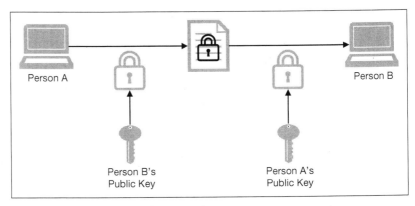

Figure 4-29. Asymmetric Encryption

decipherable only with person B's private key, which is only in person B's possession. If person B wants to send a confidential message to person A, then person B encrypts the message with person A's public key and person A must use their private key to decrypt the message. As a result, asymmetric encryption requires four keys for two-way communication, as opposed to symmetric encryption that requires only one, and so is more complex.

The most common asymmetric key algorithms are:

- **RSA** – Named for the inventors Ron Rivest, Adi Shamir, and Leonard Adleman

- **Diffie–Hellman** – Named for the inventors Whitfield Diffie and Martin Hellman

- **DSS** – Digital Signature Standard

Asymmetric key systems ensure a good security level; however, their complexity means they are slower and computationally more demanding than symmetric key encryption. Hybrid encryption systems use symmetric and asymmetric systems, combining the advantages of the two systems: the safety of the public key and the speed of the symmetric key.

In the hybrid system, a public key is used to safely share the symmetric encryption system's private key. The real message is then encrypted using that key and sent to the recipient.

Because the key-sharing method is secure, the private key used for the encryption can be changed periodically (for instance, at the start of every new session). For this reason, it is sometimes called the *session key*. If the session key is intercepted, the interceptor is only able to read the messages encrypted with that key. In order to decrypt other messages, the interceptor would have to intercept other session keys.

IPsec (short for IP Security) is a secure form of IP communications that has encryption embedded. It has two components:

1. The IPsec protocol itself, which specifies the information to be added to an IP packet and indicates how to encrypt packet data.
2. The Internet Key Exchange, which uses asymmetric key encryption.

IPsec works in two modes of operation: *transport* and *tunnel*. In transport mode, only the IP data is encrypted—not the IP headers themselves. This allows intermediate nodes to read the source and destination addresses. In tunnel mode, the entire IP packet is encrypted and inserted into another IP packet with its own header.

IPsec is often used to support secure communications over virtual private networks.

Remote Access

Poorly designed or controlled remote access to mission critical systems can provide a major security vulnerability that can be exploited by unauthorized individuals. Authorized individuals, employees or contractors, can also create problems that can result in system failure if not properly managed. Key objectives in providing secure remote access are:

- All nonessential remote access connections to systems are removed.

- All essential connections to systems are documented and approved.

- All essential connections are secured in accordance with standards and recommended practices.

- Mission critical networks are segregated from corporate and external environments.

- Remote access is limited to authorized and competent persons carrying out specific approved tasks.

It is good practice to enable and disable remote access from the mission critical facility, rather than allowing unlimited access to systems. Logging successful and unsuccessful access attempts and remote access usage and reviewing these logs are critical to the prompt detection of unauthorized access.

Cyber Hygiene

The term *cyber hygiene* covers the day-to-day tasks that are necessary to ensure the basic elements of the defense-in-depth approach are in place. The key tasks are:

- Access control
- Malware prevention and patching
- System hardening

- Removable media control
- Backup and recovery

Access Control

Access control is about knowing who can access systems and what they can do. Good cyber hygiene ensures that:

- All manufacturer default passwords are changed before equipment is operational.
- Passwords are kept confidential, are not shared, and are not posted for all to see.
- Passwords are changed regularly and are "strong" (i.e., they are not easily cracked).
- Any user who has changed roles or has left the organization has had their access removed.

Malware Prevention and Patching

Malware is constantly changing and, as a result, it is difficult to prevent any possible malware attack. However, good cyber hygiene will:

- Keep equipment up to date with the latest security updates and malware detection signatures
- Minimize the likelihood of malware being introduced into systems by carefully managing removable media access

System Hardening

System hardening ensures that equipment has all unnecessary programs and services removed or disabled so that the likelihood of using a vulnerable program or service is reduced. With good cyber hygiene, the status of equipment is carefully monitored to make certain that programs are not installed or services are not enabled unless they are necessary for the operation of the system and their use has been approved.

Removable Media Control

Removable media, such as universal serial bus (USB) drives, are a major source of malware. Removable devices are designed to transfer material from machine to machine, and unwanted material can also be inadvertently transferred in the process. This is especially true when personal devices are used, as these may be connected to home machines with limited or no anti-virus protection.

Removable media is not limited to USB drives. A CD or DVD can also contain malware. Smart phones and tablets that are charged using a USB port are also a threat, as these devices contain drives that can be infected with malware.

A comprehensive security management system will minimize the need to use removable media by implementing alternative secure communications paths for transferring data and files. However, where removable media must be used, good cyber hygiene will:

- Restrict the access to USB ports and machine drives. Port locks can be applied that although not impervious to a determined intrusion, can deter the casual user from inserting a device.

- Provide protected removable media that are secured when not in use and are reformatted and virus-checked before each use.

- Provide a means to virus-scan incoming media and transfer contents to protected removable media.

Backup and Recovery

In the event of a security incident, it is essential that systems be restored to known working order as soon as possible. Backups play a key part in this process. A good backup and recovery policy will ensure that:

- Backups are taken at a frequency commensurate with changes in the system, so that restoration does not lose a significant amount of changes. Good practice is to maintain backups of the image, containing the operating system and the applications, as well as backups of any data. This approach greatly reduces the time to restore a system in the event of an incident.

- Backups are logged, labeled, and stored in a readily accessible location, with copies in alternate locations in the event of a major disaster.

- Backups are tested regularly to ensure that they will work in the event they are required.

Detection and Prevention Systems

An intrusion detection system (IDS) monitors networks or devices for malicious activity. Networks are monitored by network-based intrusion detection systems (NIDS) and devices are monitored by host-based intrusion detection systems (HIDS). Figure 4-30 shows how IDS protects networks and devices.

An IDS uses signatures, similar to those used by anti-virus software, to detect known attacks. As with anti-virus protection, effective IDS protection requires regular updating of signatures, because only known attacks can be detected using these signatures. However, the IDS has a normal baseline for the network or device that it can use to compare current activity against to detect new, unknown attacks. However, in this

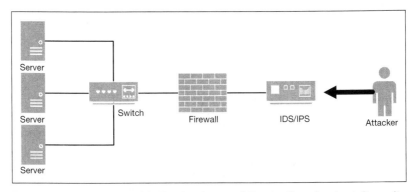

Figure 4-30. IDS and IPS Monitoring and Protecting Against Security Attacks

mode of operation, false positives are more likely. An intrusion prevention system (IPS) works with an IDS to block malicious activity when it is detected.

Care must be taken when deploying IDS and IPS technology in the process control environment. The IDS can generate significant additional traffic that may affect the operation of other networked devices. In addition, the operation of process control equipment is not well understood by these tools, so false positives and associated inhibition of functionality can occur.

Review Questions

4.1 How many layers does the Open Systems Interconnection Reference Model contain?

    A. 4

    B. 5

    C. 6

    D. 7

4.2 Which IP addressing standard uses 32 bits to represent a network address?

A. IPv1

B. IPv4

C. IPv6

D. IPv32

4.3 What is the term used to define a network that usually spans several buildings in the same city or town?

A. Local area network (LAN)

B. Metropolitan area network (MAN)

C. Wide area network (WAN)

D. Virtual private network (VPN)

4.4 Why should Wired Equivalent Privacy (WEP) not be used to secure a wireless network?

A. It uses a hard-coded default password that is easy to guess.

B. It consumes too much power, which makes it inefficient.

C. It uses a short encryption key that can be quickly cracked.

D. It is very expensive to main the encryption keys it uses.

4.5 What is the name of the network protocol that provides a secure emulation of a traditional computer terminal?

A. Secure Shell (SSH)

B. Secure Sockets Layer (SSL)

C. Secure File Transfer Protocol (SFTP)

D. IP Security (IPsec)

4.6 What is the name of the network protocol that is responsible for converting IP addresses to domain names?

A. Internet Control Message Protocol (ICMP)

B. Dynamic Host Configuration Protocol (DHCP)

C. Address Resolution Protocol (ARP)

D. Domain name system (DNS)

4.7 In a process control network, what is the name of the device that measures a physical value and converts it to a value suitable for processing by a control system?

A. Actuator

B. Sensor

C. Processor

D. Controller

4.8 What device is used in a process control network for local monitoring and control?

A. Programmable logic controller (PLC)

B. Operator interface terminal (OIT)

C. Remote terminal unit (RTU)

D. Distributed control system (DCS)

4.9 What does a trend display show in a distributed control system (DCS)?

A. Time-series data, such as flow, temperature, and pressure

B. Graphical representation of plant status, such as tank level

C. List of alarms, such as high temperature and low pressure

D. List of user actions, such as logging in and changing set points

4.10 What is a key difference between a programmable logic controller (PLC) and a remote terminal unit (RTU)?

A. A PLC is generally used to communicate over local networks, whereas an RTU is used to communicate over wide area networks.

B. A PLC is only used to communicate with a distributed control system (DCS), whereas an RTU is only used to communicate with a supervisory control and data acquisition (SCADA) system.

C. A PLC only uses Ethernet communications, whereas an RTU only uses serial communications.

D. A PLC contains local data storage for real-time data, whereas an RTU does not.

4.11 Which of the following process control network communications protocols would most likely be used to communicate with an intelligent electronic device (IED)?

A. Modbus

B. DNP3

C. EtherNet/IP

D. PROFINET

4.12 What is the name of the process control network communications protocol that is used to configure or calibrate an analog sensor using the same connection that provides the analog measurement?

A. BACnet, the Building Automation and Control Network protocol

B. CAN, the Controller Area Network protocol

C. HART, the Highway Addressable Remote Transducer protocol

D. AS-i, the Actuator Sensor Interface

4.13 What recent innovation allows a single physical server to support multiple parallel operating environments?

A. Duplication

B. Cloud computing

C. Open source

D. Virtualization

4.14 Which cybersecurity defense-in-depth protection involves management and control of system and device passwords?

A. System hardening

B. Access control

C. Malware protection

D. Intrusion detection

4.15 What are the core activities defined in the *NIST Cybersecurity Framework*?

A. Identify, secure, detect, respond, and recover

B. Plan, protect, detect, respond, and recover

C. Identify, protect, analyze, respond, and recover

D. Identify, protect, detect, respond, and recover

4.16 What type of redundant design involves running duplicate components alongside the live equipment and swapping them in manually in the event of a failure?

A. Cold standby

B. Warm standby

C. Medium standby

D. Hot standby

4.17 What type of encryption scheme involves two keys: public and private?

A. Symmetric encryption

B. Asymmetric encryption

C. Dual encryption

D. Public encryption

4.18 What is a good security practice to employ when providing remote access to mission critical systems?

A. Ensure remote access is limited to authorized and competent persons only.

B. Ensure firewalls are programmed to allow all process control communications protocols.

C. Ensure high-speed connections are used to avoid slow service to remote users.

D. Ensure remote users have direct access to key devices to avoid any delays.

4.19 How frequently should mission critical equipment be backed up?

A. All equipment should be backed up daily to minimize loss of data or changes.

B. Only critical equipment should be backed up daily to minimize loss of data or changes.

C. Equipment should be backed up so that it can be restored without losing significant data or changes.

D. Equipment should be backed up during system downtime to avoid loss of service.

4.20 What is one key reason why care must be taken when deploying an intrusion detection system (IDS) in a mission critical network?

A. IDSs may block regular activity from network devices if it is considered suspicious.

B. An IDS is old technology that is difficult to integrate with modern network devices.

C. IDSs generate a significant amount of network data that can affect the operation of other networked devices.

D. IDSs are complex to program and cannot be modified once deployed.

5
Operations

Introduction

Establishing and following sound operational procedures is essential to ensuring a safe, secure, and reliable mission critical organization. Key operational elements are:

- A comprehensive set of standard operating procedures that define how tasks are to be done

- Effective troubleshooting, repair, and restoration capabilities to minimize system downtime

- Reliable monitoring, alerting, and response tools to provide early notification of issues

- Regular audit and maintenance activities to ensure conformance with procedures, and to identify areas for improvement

- A rigorous change management process to minimize downtime and avoid the introduction of new issues

- A set of processes to manage life-cycle and resource management issues across the organization

The Standard Operating Procedure

A standard operating procedure (SOP) is a set of step-by-step instructions designed to help workers complete routine operations or activities. Examples of routine activities that would normally have an associated SOP are:

- System backup
- Device calibration
- Equipment inspection

Organizations create SOPs for activities to:

- Minimize the opportunity for error, by prescribing the method and leaving no scope for interpretation
- Improve consistency and reliability, by creating a well-defined procedure that can be followed repeatedly with the same outcome
- Improve operational efficiency, by identifying the optimal procedure
- Achieve and document compliance to regulatory requirements

Whereas international, national, industry, and company standards define *what* must be done, SOPs define *how* it is done. Figure 5-1 shows the relationship between standards and SOPs.

SOP Version Control

SOPs are living documents, subject to revision for a variety of reasons, including:

- To correct errors in the procedure
- To add further detail to clarify the procedure

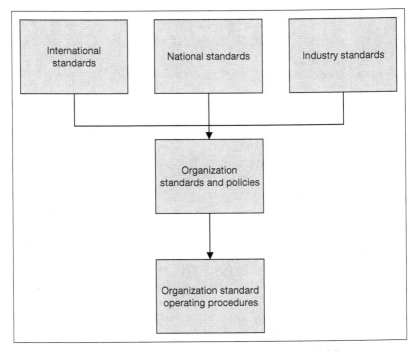

Figure 5-1. The Relationship Between Standards and SOPs

- To revise the procedure after identifying process improvements or efficiencies

Because multiple versions of an SOP may exist within an organization, it is essential that good document version control be in place. Key aspects of good SOP version control are:

- Avoid situations in which employees work from paper copies of SOPs, as there is no way of verifying that a paper copy is the latest version. Typically, paper copies are marked clearly to indicate that they must not be used without verification.

- Provide a single source of SOP version information, so that it is clear which version's the latest and what changes have been made from previous versions.

Many organizations use electronic document management systems to streamline and automate document version control and storage. These systems may provide features such as:

- Centralized storage of documents with search capabilities
- Automated update of document version after edits
- Automated notification to document distribution members to review and approve changes

The Elements of an SOP

Each organization will have its own format and style for SOPs. However, all SOPs should contain the content shown in Table 5-1, as a minimum.

Table 5-1. Essential Elements of an SOP

Document identification	At minimum, an SOP must have a title, reference number, and version number.
Version history	A table showing revisions to the SOP with date, author, and summary of changes.
Purpose, scope, and objectives	Defines what the SOP is intended to cover. It also defines what the SOP does not cover, if applicable.
Definitions	A table of definitions of key words, acronyms, and abbreviations.
Assumptions, dependencies, and constraints	A list of points that must be addressed or in place before the SOP can be executed. For example, details of systems that must be running during the SOP, notification of SOP execution, employee training, etc.
Personnel competencies	A definition of the necessary skills and knowledge to undertake the work.
Equipment and supplies	A list of tools, equipment, or materials that will be required to execute the SOP.
Process description	May be presented in a variety of formats, based on the application. For instance, in some cases, a numbered list of steps is suitable, whereas a flowchart may be more suitable in others. Depending on the procedure, it may be necessary to identify who performs which step(s).

Examples of SOP Presentation

SOPs may be presented in various forms. Figures 5-2 and 5-3 show a simple process description involving three roles, in tabular and flowchart form.

Separation of Duties

Separation of duties is an essential element of mission critical operations. It involves ensuring that more than one person is required to complete a particular task where safety or security is paramount so that:

- The risk of human error is reduced
- The risk of fraud is reduced

Typical separation of duties may involve:

- Separate electronic authorization for particular actions, such as to change set points in a control system or to migrate software among development, testing, and production environments

Step	Description	Role A	Role B	Role C
1	Remove backup media from storage facility	X		
2	Insert backup media into drive X		X	
3	Execute backup program		X	
4a	On successful completion, remove backup media from drive X		X	
4b	If unsuccessful, report to Role A to resolve reported issue		X	
5	Return backup media to storage facility	X		
6	Update backup log			X

Figure 5-2. A Simple Process Description in Tabular Form

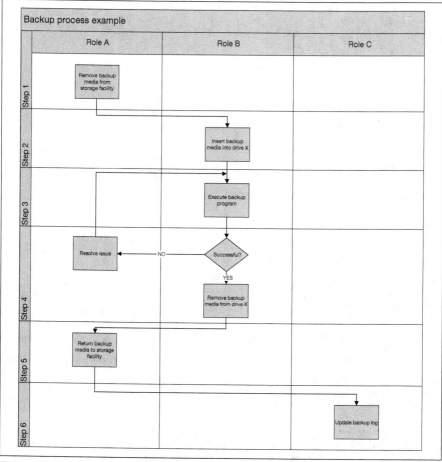

Figure 5-3. A Simple Process Description in Flowchart Form

- The use of multiple security keys, held by separate personnel

Where it is not possible or practical to separate duties (e.g., in very small organizations where skilled personnel are limited), alternate controls should be in place, such as:

- Audit trails to track who took what action and when

- Periodic supervisory reviews to verify all tasks are being performed as expected

Troubleshooting, Repair, and Restoration

Mission critical systems, by definition, must operate without unplanned interruption to avoid an impact on the organization concerned. However, it is unrealistic to expect that system components, hardware or software, will never fail. Effective troubleshooting, with subsequent repair and restoration capabilities, is essential in a mission critical environment.

Performance Objectives

There are many measures related to the performance of a mission critical system. Organizations will utilize one or more to define their expectations and this should translate into requirements relating to troubleshooting, repair, and restoration.

Availability is the most common measure of system operation. Availability is the probability that the system is operating properly when it is required. Availability is measured as a percentage over a defined period of time (e.g., per day, per month, or per year). Availability over a year is the most frequently used measure. Table 5-2 shows availability percentages and the corresponding amount of downtime over the period of 1 year.

Another measure often confused with availability is *reliability*. These terms are not the same. Reliability is a measure of the probability of a component or system to perform a required

Table 5-2. System Availability and Corresponding Downtime over 1 Year

90%	99%	99.9%	99.99%	99.999%	99.9999%
40 days	4 days	9 hours	50 minutes	5 minutes	30 seconds

It is typical for availability to be quoted as "nines" (e.g., "five nines," which means 99.999%, or a downtime of 5 minutes in 1 year).

function for a specific period of time without failure, in a specific environment. Unlike availability, reliability does not account for any repair actions that may take place.

There is a third term, *maintainability*, which connects reliability with availability. Maintainability is a measure of the ease with which a product can be maintained. Availability is the combination of reliability and maintainability, so it is important to note that high reliability components (or systems) alone do not result in high availability systems. Maintainability is an essential element for successful operations.

The relationship between maintainability and availability is seen in practice with the definition of maintenance windows by the operations and maintenance staff. A maintenance window is a defined period of time during which planned outages and changes to systems or services may occur. By defining standard maintenance windows, potentially impacted users and activities can prepare for the possible disruption. Oftentimes with software updates, the maintenance window is in the middle of the night, when the fewest users are active. In some systems with planned redundancy, a backup resource is put into operation while the main resource is being serviced. After the main resource is serviced, it is tested and returned to service. The process is repeated with the backup resource.

Table 5-3 lists some other common measures in mission critical operations.

Troubleshooting Principles

There is no precise method for troubleshooting. A failure can be diagnosed and rectified in many different ways. However, there are some general principles that should be adopted to create a more structured approach to troubleshooting. Figure 5-4 shows the key steps in troubleshooting.

Table 5-3. Common Performance Measures in Mission Critical Operations

Key performance indicator (KPI)	A measure that is essential to identifying the status of operations in an organization. The actual KPI(s) will vary depending on the organization. For example, a manufacturing company might identify the number of items produced per day, whereas a water company might identify the volume of water treated per day.
Overall equipment effectiveness (OEE)	Used particularly in manufacturing organizations, this measure combines quality, performance, and availability to give an overall score as a percentage. An OEE of 100% would indicate the organization is producing no defective parts, as fast as required, and there is no loss of service.
Recovery time objective (RTO)	The duration of time and a service level within which a process must be restored after a disaster (or disruption) in order to avoid unacceptable consequences associated with a break in service.
Recovery point objective (RPO)	The acceptable amount of data loss measured in time (e.g., data must be restored within 2 hours of a disaster for the loss of that data to be acceptable).
Service level agreement (SLA)	A measure agreed between two or more parties for the operation of a system or service. As with KPIs, actual SLAs will vary depending on the organization and the overall agreement itself.

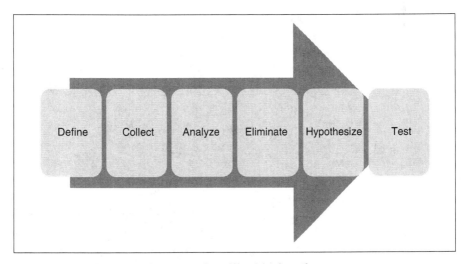

Figure 5-4. A Structured Approach to Troubleshooting

Troubleshooting begins with a report of a problem. The first stage of troubleshooting is to *define* the problem in terms that will allow the rest of the troubleshooting process to continue along the correct path. A badly defined problem can result in time and energy being wasted looking in the wrong areas.

The next step is to *collect* data about the problem. This may include reports from users; network or equipment logs; and possibly even outside data (such as weather reports). Once data is collected it can be *analyzed*. The purpose of analysis is to use the data, with knowledge of the system behavior, to identify possible causes.

Once possible causes are identified, some can be **eliminated**, so that the likely cause is *hypothesized*. The hypothesis can be *tested*, often by implementing a particular solution. If it is not practical to implement a particular solution at this stage (e.g., because it will further disrupt operations), then it might be necessary to collect further data to analyze.

Critical Repairs

The structured approach to troubleshooting may appear long-winded, but it provides the best chance of resolving issues in a timely and reliable manner.

In some situations, there may be a critical need to restore operations without fully troubleshooting the cause. For example, a device may be replaced with a spare before ascertaining the cause of the failure of the original device.

In this case, it is important that troubleshooting is performed after the restoration, to avoid similar failures in the future. For example, if a device failed because of an external power supply fault, then the external power supply must be replaced if the issue is to be avoided in the future.

Monitoring, Alerting, and Response

Logging and Monitoring

Mission critical organizations will use a variety of logging and monitoring solutions to track the status of networks and equipment:

- *Industrial control systems*, such as supervisory control and data acquisition (SCADA) systems; distributed control systems (DCSs); heating, ventilation, and air conditioning (HVAC) systems; and building automation systems (BASs), will be used to monitor the status of operational equipment, such as plant, machinery, and buildings. Although they have varied names, the basic function and operation of these systems is the same. They collect data from sensors and report this data, along with alarms to highlight areas of concern, to operators. Other infrastructure equipment, such as uninterruptible power supplies (UPSs), may report data and alarms through these systems or directly to operators through specialist applications.

- *IT monitoring systems* are used to collect and report data on IT equipment, such as servers, workstations, and databases. Such monitoring can provide data about system performance, memory usage, disk space, processor usage, and the like. Windows Management Instrumentation (WMI) is a Microsoft tool that can provide such information. Operating system event logs can also be analyzed to identify events that occur on a system, such as data loss, application performance, failed logons, and attempts to access secure files.

- *Network monitoring systems* are used to collect and report data on network equipment, such as routers and switches. In modern networks, devices support protocols such as the Simple Network Management Protocol (SNMP). SNMP allows the exchange of information between devices in a network and the network management system (NMS). In SNMP, devices contain an SNMP agent that is responsible for communicating with the NMS. A management information base (MIB) is a database that defines what information the device collects. All SNMP-cable devices support a basic level of interaction with SNMP-based NMSs, but more advanced devices may support more features.

Events, Alarms, Alerts, and Notifications

Events, alarms, alerts, and notifications are terms describing similar outputs of monitoring systems. Although vendors and users may define them in different ways, in general they are defined as in Table 5-4.

Alarm Priorities

In most systems, alarms will be defined with different priority or severity levels. This allows operators to quickly determine where action is required first. Usually, alarm prioritization is

Table 5-4. Events, Alarms, Alerts, and Notifications

Event	This indicates when something happens in the system that is important to record but does not require immediate action; for example, if a pump starts or stops or when an authorized user logs into a system.
Alarm	This indicates when something happens that requires action; for example, high pressure, low temperature, or an unauthorized attempt to log in to a system.
Alert	In some systems, an alert is an umbrella term for all events and alarms. In other systems, an alert is another term for alarm or event.
Notification	This is usually an alternative term for an event (i.e., an indication that something important occurred but does not require immediate action).

configurable by the organization, as each organization has different priorities. Alarms may be prioritized using a variety of means:

- Numbering from 1 to 10, where 1 is most critical and 10 is least critical (or vice versa)

- Lettering from A to Z, where A is most critical and Z is least critical (or vice versa)

- Other means, such as a three-level categorization of "normal, warning, and alert" or "minor, serious, and critical"

Alarm Handling and Escalation

Alarms are presented to operators via workstation applications. An example of a typical SCADA alarm display is shown in Figure 5-5.

The main objective of the alarm facility is to report issues to operators, allow the operators to analyze them

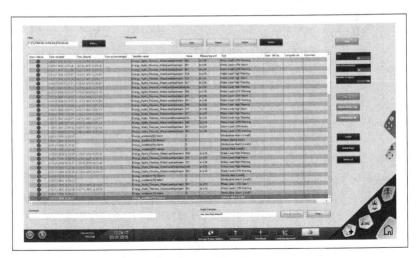

Figure 5-5. A Typical SCADA Alarm Display

Source: Alarm screen in zenon, courtesy of COPA-DATA.

(by providing additional data as necessary), and then make the correct action determinations (e.g., send a technician to replace a component).

Most alarm systems operate on an alarm life cycle. This life cycle ensures alarms are handled in a reliable manner. For example, in a SCADA system, an alarm may exist in one of the states shown in Table 5-5.

Systems will provide various means to help the operators assess the alarms, such as the use of colors and/or symbols to indicate status (e.g., acknowledged and unacknowledged).

Many systems will have multiple levels of supervision. This is to ensure that critical alarms, especially time-dependent ones, are handled as required. If the primary operator fails to deal with an alarm in a particular timescale, the system may be configured to escalate the alarm to a supervisor.

Newer monitoring systems also offer configurable methods to deliver alarms as text messages, emails, voice calls, and Hypertext Transfer Protocol (HTTP) GET/POST. These methods

Table 5-5. Alarm States

Unacknowledged	Typically, this is a new alarm that has not been processed by the operator.
Acknowledged, but not cleared	The operator has acknowledged the alarm and may have dispatched someone to rectify the situation, but it is not yet completed.
Cleared, but not acknowledged	The alarm occurred and then cleared (e.g., a tank experienced a high level but it subsequently reduced to normal), but it remains in the system so that the operator acknowledges it.
Acknowledged and cleared	The operator acknowledged the alarm and the underlying cause has been rectified.

Other states, such as deleted and timed-out, may also exist, depending on the system.

can be used in conjunction with escalation options to keep trying various routes to ensure the alarm is addressed.

Event–Alarm Correlation

Operators in mission critical organizations will often correlate event and alarm data from one or more systems to determine actions. For example:

- From a security perspective, it might be important to compare the records in a secure card entry system with records of failed logins to servers or workstations.

- In a SCADA system, it might be necessary to analyze pump start and stop events to correlate with a high tank level alarm.

False Positives and False Negatives

Monitoring systems can fail to report alarms or can report alarms when no incident is occurring. The event matrix shown in Table 5-6 highlights the possible situations that can occur. A true-positive or true-negative is desired: that is, when an incident is occurring, a corresponding alarm is received (true-positive); or when there is no incident, and no alarm is received (true-negative).

Table 5-6. Event Matrix

Alarm Received	Incident Occurring	
	True	False
Positive	True-positive (Alarm received, incident occurring)	False-positive (Alarm received, no incident occurring)
Negative	True-negative (No alarm received, no incident occurring)	False-negative (No alarm received, incident occurring)

In some cases, false-positive and false-negative situations can occur. In a false-positive situation, there is no incident occurring but the system generates an alarm. In a false-negative situation, an incident is occurring but no alarm is reported.

Both false-positive and false-negative situations can happen for a variety of reasons, including:

- **Incorrect system configuration** – For instance, the level of the tank, or the server limit being monitored, is incorrect.

- **Faulty sensor** – For instance, the sensor reports nothing or the wrong value.

Many system components now provide diagnostic data, such as a general "heartbeat" to indicate that the device is active, and to indicate the quality of readings, to aid in alarm analysis.

Both false-positive and false-negative situations are cause for concern. False-positive situations create a loss of confidence in monitoring systems, which may result in a real situation being ignored. False-negative situations mean that real incidents are being missed. It is essential that alarm-monitoring systems be thoroughly tested to exercise all possible conditions.

Audit and Maintenance

Keeping a mission critical operation running smoothly requires constant attention to the systems involved.

Regular, preemptive maintenance is essential to avoid unexpected downtime and the subsequent issues it can create. Periodic audits, involving inspection or examination of systems, are also required to ensure compliance to requirements. Changes to systems must be carefully managed, using a rigorous process, to avoid unforeseen issues.

Configuration Management

Before any effective audit and maintenance regime can begin, it is essential to have a complete inventory of all system hardware and software. This process is often called configuration management (CM).

The inventory is sometimes called an *asset register* or *configuration management database*. It can be as simple as an Excel spreadsheet or can be a purpose-made relational database and application.

At minimum, a system inventory should contain the following for each component (called configuration items—CIs in CM-terminology) in a system:

- A unique identification number (to make tracking easier, a label with this number should be affixed to the device)
- The manufacturer's make and model number
- The device serial number
- A brief description of the device (e.g., pump control PLC or Operator Workstation #1)
- The location of the device
- The version number of the device hardware
- A list of all software installed on the device and all associated version numbers
- Any address information (e.g., IP address or protocol reference)

In some configuration management databases, it may be possible to store configuration and program files for the device so they can be controlled from the same location.

The system inventory should be constantly maintained to ensure it is kept up to date with changes, such as replacement hardware, updated software, and changes to addressing. The importance of the CM process is usually recognized in software environments, such as ensuring that the validation environment and the production environment are both running the same version of software. However, physical aspects must also be under CM, to avoid such problems as trying to connect a 6-inch pipe to a 12-inch pipe or a 110-volt instrument to a 220-volt circuit.

Upgrade Management

Upgrade management of mission critical systems must be performed with extreme care. A failure to update a device may leave the system open to a failure, such as due to a known software issue, or a cyber incident due to a known cyber vulnerability. However, a poorly managed or untested update to a device can create an unplanned outage.

In order to minimize the likelihood of unplanned outages, mission critical organizations will operate a change management process that reviews and validates all changes, including upgrades to hardware and software, and determines a methodical plan to perform the upgrade. Change management is covered later in this chapter.

Any upgrades to mission critical system components must be reflected in the system inventory.

Periodic Testing

Some devices or systems in mission critical organizations are designed to only operate when a failure condition exists, such as high-pressure in a pipeline. Periodic testing is required to exercise such devices or systems to verify that they will operate correctly when required. Depending on the device or system, it may not be possible to test until there is a plant shutdown.

Auditing

Regular audits of mission critical systems are essential to ensure that:

- **The system inventory is correct** – Although procedures may require personnel to update the inventory after a change, it is possible for some updates to be missed or to be made incorrectly. In large systems with many components, it is common to perform a sample inspection of the inventory, checking a percentage of the entries.

- **No unapproved changes have been made** – Although mission critical organizations will have a change management procedure, it is possible for this to be bypassed. For instance, if the change is perceived as trivial or if the change was made during a system failure where the focus was on returning the system to normal operation.

- **Electronic and physical access to the system is restricted to only those who need access** – Personnel who leave an organization or move to a new role are often not removed from physical or electronic access control systems, such as electronic key-pass systems or electronic account-authorization systems.

- **All periodic activities are being performed** – Periodic testing and maintenance activities will be defined for each system and these may not be performed to schedule, for instance, due to limited resources.

Personnel who are independent of the system being audited will normally undertake audits. Findings from the audit should be documented and a follow-up scheduled to verify that actions arising have been addressed.

Change Management

Change management (sometimes called Management of Change—MOC) is the most important operating procedure in a mission critical organization. Mission critical organizations, by their very nature, require safe, reliable, and stable operations. It is inevitable that changes will need to be made to mission critical systems, but they must be made in a safe and reliable manner to avoid:

- Harm to individuals
- Damage to equipment
- Harm to the environment
- Loss of service

Key aspects of an effective change management process are:

- Prior approval for work by all relevant stakeholders
- Prior notification of change to all relevant stakeholders
- Identification of all potential impacts of the change, positive and negative
- Identification of all potential hazards during the change process, and associated mitigations
- Agreement on the method for implementing the change, commonly known as the *method statement*
- Identification of the individuals approved to perform the change

In mission critical organizations, changes, no matter how small, must never be made without following the change management process.

The Change Process

Changes may be required for many reasons, such as:

- Replace obsolete equipment
- Improve a process
- Enhance monitoring of process equipment

A justification for the change is normally the starting point in the process. This justification will explain why the change is required, including the potential benefits. In addition to the benefits, it will be necessary to identify all negative impacts of the change, such as increased maintenance costs.

Once a full analysis of the pros and cons of the change is produced, a plan for implementing the change is required. This plan will identify the likely timescale for the change and the resources required; and it will define the exact, step-by-step change process (the method statement). The change process must identify a way to "back out" of the change (i.e., revert to the current situation, pre-change, if something should go wrong). Without such a strategy, a mission critical organization can experience a significant unplanned outage.

All relevant stakeholders (operators, maintenance technicians, managers, and executives) must review and approve the change documentation before any work can be undertaken. Many organizations define who the relevant stakeholders are by producing a RACI matrix—as in Responsible (R), Accountable (A), Consulted (C), and Informed (I) (see Figure 5-6). Some organizations use a modified version of this matrix, called RASCI,

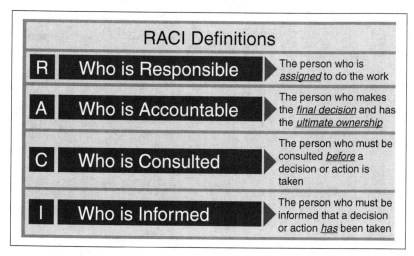

Figure 5-6. RACI Matrix Definitions

where the S identifies the Stakeholders responsible for the support of the system or process, or anyone else who is needed to complete the task.

Implementing the Change

Once the relevant stakeholders approve a change, it can be implemented.

Before any work can be performed, a permit to work (PTW) is required. The PTW is an essential safety management control. It defines:

- The work that will be done
- Where the work will be done
- When the work will be done
- The persons doing the work

The PTW process involves assessing the hazards involved in the work and identifying any precautions that are required to control these hazards.

A responsible person will assess the work and check safety at each stage. Those doing the work sign the PTW to show that they understand the necessary risks and precautions.

As well as managing safety of the personnel involved, the PTW process is a means of communication between management, supervisors, and operators, and those who carry out the work. It is also a means of coordinating different work activities to avoid conflicts.

"Toolbox talks" will also be used during change work. Toolbox talks are informal safety meetings that focus on safety topics related to the specific job, such as workplace hazards and safe work practices. Meetings are conducted at the location of the work, prior to the commencement of the change. Toolbox talks cover last-minute safety checks and exchange information with the change workers.

Verifying and Finalizing the Change

Once the change is implemented, it will be tested against an agreed-on test specification. There are two distinct aspects to testing:

- **Validation** – This ensures that the change meets the needs of the stakeholders, irrespective of the requirements.

- **Verification** – This ensures that the change complies with all regulations, requirements, and specifications.

To ensure continued effective support and maintenance, it is essential that all relevant documentation be updated to reflect the change. This includes any changes made during the change process.

Normally, *red line drawings* are produced during the change implementation process. These drawings (or documents in some cases) reflect changes made during the change (e.g., to correct an error in the original drawings). Once red line drawings are reviewed and approved, a final set of documents called the *as-built drawings* are produced to reflect these amendments.

Life-Cycle Management

Life-cycle management is a term to define all the activities involved in the day-to-day operation of a mission critical organization. *Life cycle* may apply to the product produced by the organization or to the assets used by the organization to produce or deliver the product.

The precise details of the life cycle of a mission critical organization will vary based on the organization; however, there are some general points that should be considered in all cases:

- Forecasting and provisioning
- Commissioning
- Support
- Obsolescence planning
- Decommissioning and disposal

Forecasting and Provisioning

For a mission critical organization, effective forecasting and provisioning is essential. Organizations that involve manufacturing require raw materials to be available as and when they are needed. A failure to accurately forecast or provision can result in significant downtime and associated cost impacts.

Forecasting is based on an analysis of trends and requires accurate inventory details for it to be effective. Some organizations use statistical forecasting, based on past demand, and some use nonstatistical forecasting, where planners use subjective factors to determine future demand.

Regardless of the forecasting method, the end result is to determine what raw materials are required and when. If under-provisioned, production downtime can be experienced. If over-provisioned, there can be wastage. A whole discipline, called *lean manufacturing* (or *just-in-time manufacturing*), has been created to identify techniques and tools that can improve provisioning to the point where raw materials are available just when they are needed.

Kanban (derived from the Japanese for billboard) is a scheduling system originally developed by Toyota and now widely used in manufacturing. Kanban allows manufacturers to maintain low inventory levels by tracking consumption and signalling when new orders are required to replenish stock, while maintaining a continuous production process.

Commissioning

When new a plant is deployed, or an existing plant is upgraded, a formal process is conducted to ensure that the plant is ready for operational use. Commissioning begins immediately after installation. The key elements of commissioning are:

- Inspection of the plant and associated instrumentation and power supplies. This is conducted before any equipment is energized. Normally, a commissioning team, independent of the installation team, visually inspects the installation against the relevant drawings. Wiring continuity checks are usually performed at this stage.
- Calibration or setup of sensors and actuators.
- Testing the plant in its target environment after installation. Normally, the vendor (or primary contractor)

will produce a test specification that exercises the plant against its agreed-on specification. The end user will normally witness this testing.

- Testing the safety features of the plant, such as emergency shutdown controls.

Support

Once the plant is commissioned, it enters normal service. Until the plant is decommissioned, it must be supported to ensure that the mission critical operation is not adversely impacted.

Plant failure rates will vary considerably depending on environment, application, and the quality and frequency of maintenance. In general, the "bathtub" curve is used to present the typical failure rate of plant over its life cycle (see Figure 5-7). During the early part of the plant's life cycle, failures are more frequent. Possible causes might be incorrect installation or poor quality manufacturing. After a period of time, the plant enters its "useful-life" period, in which failure rates are typically stable. Finally, toward the end of life for the plant, the failure rate begins to increase (e.g., as components wear out).

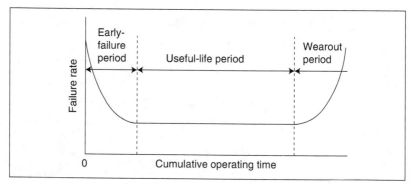

Figure 5-7. The "Bathtub" Curve, Showing Failure Rates over an Entire Life Cycle

Traditional plant maintenance is based on the preventative maintenance approach. In this approach, maintenance is performed at regular intervals, based on prior knowledge of the failure frequency. The aim is to perform maintenance before a failure is likely to occur. The problem with preventative maintenance is that no two installations are the same. For example, one item may operate more often than another identical item and as a result:

- Unplanned outages occur, before routine maintenance can take place, on equipment that experiences more usage.

- Resources are misdirected to perform unnecessary maintenance on equipment that is used less often.

The predictive maintenance approach relies on the actual condition of equipment to determine when maintenance is required. In order for predictive maintenance to work, condition monitoring is required.

Condition monitoring is the process of monitoring aspects of equipment operation, such as vibration or temperature, and identifying changes that indicate a potential failure is imminent. For example, if vibration readings on rotating equipment increase over time, this might suggest that parts are loosening and may eventually come apart. Organizations will typically use two independent methods for identifying failures to avoid false-positive reports. For example, a decline in flow rate and an increase in impeller vibrations would be used to identify an impending failure of a pump.

It is possible to identify trends and prepare for them through continuous monitoring. Scheduling resources and ordering spare parts for a planned outage allows for repairs to be performed before an unplanned outage occurs.

Obsolescence Planning

Obsolescence planning is needed to prepare mission critical organizations for changes that will be required in the future. Obsolescence occurs because of technological evolution, market forces, and legislation. For example:

- Vendors may be unable to support their products because of unavailability of components.

- Communications technology changes may require products to be replaced.

- Equipment may not meet new or upcoming regulations, such as environmental requirements.

In mission critical operations, it can take months or even years to prepare for equipment replacement. A lack of obsolescence planning can result in the need to make changes quickly, and this should never be done in mission critical organizations. As already discussed, change management is a critical process that must always be performed to avoid unplanned outages or unexpected failures.

Decommissioning and Disposal

The plants or equipment are decommissioned at the end of their useful life or when the cost of operating or maintaining them becomes too high.

Decommissioning will normally form part of the replacement program. In mission critical organizations, scheduled outages are called *plant turnaround, maintenance turnaround, plant shutdown,* or *plant outage.* The whole plant is stopped for an extended period of time that may range from days to weeks, depending on the scale of the operation.

Once a plant is decommissioned, it will need to be disposed of in a suitable manner. Note that:

- Some plants may contain hazardous materials (such as nuclear waste and toxic chemicals) that require special disposal techniques.

- Computerized equipment may contain sensitive information (such as the organization's intellectual property, personal identification information, usernames, and passwords) and must be disposed of carefully to avoid this information falling into the wrong hands.

Resource Management

Resource management is required to ensure that mission critical organizations have the supplies, equipment, people, facilities, and services they need to operate, both under normal circumstances and in emergency situations.

In normal circumstances, the objective is to achieve 100% utilization of resources, so that there is no waste. One method to achieve this is called *resource leveling*. The object of resource leveling is to find the optimum balance to achieve a particular outcome, such as minimizing costs or achieving a particular service level.

In addition to planning for normal circumstances, the organization must ensure that resources are available for emergency situations. The organization will need to consider all possible emergency scenarios and ensure that resources are in place to address them. Typical considerations for such situations are:

- Spare parts available for deployment to meet the required service objectives
- Personnel available to be called in to supplement on-duty staff
- Plans in place with local, state, and federal emergency services in the case of particular emergencies

Review Questions

5.1 What is one of the primary purposes for standard operating procedures (SOPs)?

A. To minimize the opportunity for error when undertaking an activity

B. To define what must be done when undertaking an activity

C. To define why an activity is to be undertaken

D. To define the regulations relating to an activity

5.2 What is one good practice that should be employed to maintain good version control of SOPs?

A. Issue printed versions of SOPs to all users to allow them to work from when on site.

B. Issue editable versions of SOPs to all users to allow them to mark up their own changes.

C. Provide a single source of SOP version information to ensure that users know which version of an SOP is to be used.

D. Combine all SOPs into a single document to make it easier to maintain.

5.3 What is the common term for a good practice to reduce the risk of fraud or human error?

A. Segregation of roles

B. Separation of duties

C. Division of responsibilities

D. Management of change

5.4 What does an availability of "five nines" correspond to in terms of downtime in a single year?

A. 4 days

B. 9 hours

C. 50 minutes

D. 5 minutes

5.5 What is the name of the measure that combines quality, performance, and availability to give a single overall score?

A. Service level agreement (SLA)

B. Overall equipment effectiveness (OEE)

C. Key performance indicator (KPI)

D. Recovery time objective (RTO)

5.6 What is the name of the protocol that is most often used to collect network management data?

A. Simple Network Management Protocol (SNMP)

B. Supervisory control and data acquisition (SCADA)

C. Network management system (NMS)

D. Distributed control system (DCS)

5.7 What is the key difference between an event and an alarm?

A. Unlike an alarm, an event is something that has occurred that requires no immediate action.

B. Unlike an event, an alarm results in an audible signal to notify operators.

C. Unlike an alarm, an event is not logged by the relevant monitoring system.

D. Unlike an event, an alarm is used to notify operators of security issues.

5.8 In the context of alarm management, what is a false positive?

A. An alarm was received and an incident is occurring.

B. An alarm was received but there is no incident occurring.

C. An alarm was not received but an incident is occurring.

D. An alarm was not received and there is no incident occurring.

5.9 In the RACI matrix, which defines stakeholders in a change management process, what does the abbreviation RACI mean?

A. Responsible, Accountable, Consulted, and Instructed

B. Reporting, Accountable, Consulted, and Informed

C. Responsible, Accountable, Consulted, and Informed

D. Reporting, Approved, Consulted, and Informed

5.10 What is the name of the curve that shows typical equipment failure rates over an entire lifecycle?

A. Bathtub curve

B. Reliability curve

C. Probability curve

D. Failure curve

6
Safety and Physical Security

Occupational Safety

Personal Protective Equipment

Personal protective equipment (PPE) is equipment worn by personnel to minimize their exposure to workplace hazards, such as:

- Falling objects
- Hazardous chemicals
- Shrapnel or other foreign bodies
- Loud noise

PPE varies depending on the industry; however, most mission critical sectors will require some or all of the items shown in Table 6-1. Additional items of PPE may be required for specific roles or tasks. The mission critical organization will define what PPE is required and this will be strictly enforced. The organization will also define rules for inspection and replacement of PPE, based on wear and age.

Table 6-1. Typical Items of Personal Protective Equipment (PPE)

Protection for	Description	Example
Respiratory system	A mask protects against inhalation of dangerous or noxious gases or fumes. The exact design of the mask will vary, depending on the particular hazard.	
Head	A hard hat protects the head against falling objects as well as collisions with low scaffolding or other fixtures.	
Eyes	Goggles or glasses protect the eyes against foreign bodies, shrapnel, or hazardous chemicals. Various designs exist and specific forms will be required when dealing with particular hazards. Some areas of work may require goggles over glasses to provide additional protection, and some areas may require only glasses.	
Hearing	Earplugs protect the eardrums from damage caused by loud noises found in many industrial environments (e.g., from rotating machinery). Various designs are in use, including over-ear versions that attach to hard hats.	

Table 6-1. Typical Items of Personal Protective Equipment (PPE) *(Continued)*

Protection for	Description	Example
Hands	Gloves protect against chemicals, contamination, and infection as well as temperature extremes and mechanical hazards. Rubber gloves provide insulation when working with electricity.	
Feet	Safety footwear comes with a protective toecap to protect against falling items. Various types of safety footwear exist when working with specific hazards. Examples include: • Footwear with sole puncture protection, where sharp objects (such as nails) are present • Electric-shock protective footwear, when working with live electrical conductors • Footwear with additional metatarsal protection, where heavy objects are present that could damage the metatarsal region of the foot	
Body	Overalls protect the wearer from various hazards, such as chemical or other spills. Overalls are often water-resistant and fireproof. They also usually include high-visibility markers to aid in locating personnel in low-light working conditions. For certain environments, special overalls may be required that provide additional levels of protection.	

Lockout/Tagout Procedures

Lockout/tagout (LOTO) is a safety procedure designed to ensure that dangerous equipment is disabled and not able to be energized during maintenance activities. It is common in many mission critical organizations, but especially those involving any form of energy, such as:

- Electricity
- Compressed air
- Gas
- Steam
- Hydraulics

The LOTO procedure requires that energy sources are isolated and rendered inoperative before work commences. This is achieved by locking a tag on the equipment, preventing it from being operated. An example of a LOTO tag is shown in Figure 6-1.

Figure 6-1. An Example of a LOTO Tag

LOTO procedures will vary from organization to organization, but the recommended practice is to affix a lock and a warning label that shows the name of the person who is performing the work. This person should be the only one holding the key to unlock the tag.

LOTO tags have multiple holes to allow for two or more people to work on the equipment (or related equipment) at the same time. In this case, each person fixes his or her own lock to the tag. In this way, the equipment cannot be activated until all workers have removed their locks.

In many organizations, industries, or countries, locks may have a particular designation indicated by color, shape, or size. For instance, the U.S. Occupational Safety and Health Administration (OSHA) recommends the following designations:

- "DANGER" – Red, or predominantly red, with lettering or symbols in a contrasting color

- "CAUTION" – Yellow, or predominantly yellow, with lettering or symbols in a contrasting color

- "WARNING" – Orange, or predominantly orange, with lettering or symbols in a contrasting color

- "BIOLOGICAL HAZARD" – Fluorescent orange or orange-red, or predominantly so, with lettering or symbols in a contrasting color

Safety Orientation and Hazardous Operations

Safety orientation is a mandatory element for employees working in mission critical organizations. The objective is to reduce the risk of potential injuries and accidents.

Safety orientation provides employees with information about general workplace hazards and those hazards that are specific to their job. Examples of aspects covered during safety orientation are:

- Working with electricity or other forms of energy
- Working at heights
- Working below the ground surface (e.g., ditches)
- Working in confined spaces
- Working with hazardous materials (e.g., H_2S gas)
- Fire safety
- Safe driving

In many mission critical sectors, regulatory requirements mandate that safety orientation be provided. Irrespective of regulations, organizations have an ethical obligation to protect employees from harm.

Organizations must keep records of safety orientation training filed in the employee's personnel file along with all his or her training records. OSHA regulations require that such records are kept for at least 5 years and are available on request.

Job Safety Analysis

A job safety analysis (JSA) or job hazard analysis (JHA) is a process to identify potential hazards that exist in that job or task. The objective is to determine appropriate procedures, safety equipment, and clothing that will be required to remove the identified hazards or to introduce safeguards to reduce the potential impact of them occurring.

In order to identify the potential hazards, it is essential that the job or task be broken down into discrete steps. For each step, the reviewers must identify what can go wrong; and for each possible event, they must consider:

- What are the consequences of the event?
- What ways could the event occur?
- What factors would contribute to the event occurring?
- What is the likelihood of the event occurring?

Job safety or job hazard analysis results must be recorded in a standard format that identifies:

- Task description
- Task location
- Identified hazards
- Identified controls
- Name of person who completed the analysis and date of the analysis

A simple example job hazard analysis based on changing a tire is shown in Figure 6-2.

Safety Data Sheets

Chemical manufacturers, distributors, or importers are required to provide Safety Data Sheets (SDSs, formerly known as Material Safety Data Sheets or MSDSs) to provide details of the hazards of their products. Employers are required to ensure that SDSs are readily accessible to all employees.

OSHA's Hazard Communication Standard (HCS) defines the details to be provided on SDSs as shown in Table 6-2.

Job: Changing a Tire
Department or location: Road

Task or Step	Hazards	Controls	Personal Protective Equipment (PPE)
Park vehicle in safe spot	1. Exposure to passing traffic 2. Fall on uneven ground	1. Park in area well clear of passing traffic. 2. Use emergency flashers. 3. Select a firm, level area.	
Remove spare wheel and tools	1. Injury from lifting heavy objects.	1. Orient spare wheel in upright position; Stand as close as possible to wheel; Bend knees slightly and lift wheel carefully and slowly; Roll to corner with flat.	Gloves
Remove hub cap and loosen wheel nuts	1. Injury from slipping wrench 2. Injury from dropped or flying object (e.g. hub cap)	1. Pry off hub cap slowly, using steady pressure. 2. Use correct wrench. 3. Apply steady pressure to wrench.	Gloves
Insert jack and raise corner with flat	1. Injury on hand jammed between jack and vehicle base. 2. Injury due to incorrectly aligned or inserted jack.	1. Check that jack is securely positioned on vehicle base and on a flat surface before raising. 2. Keep hands on jack handle or away from vehicle base while raising.	Gloves
Remove wheel nuts and take off wheel	1. Injury from wheel dropping on foot or hand	1. Maintain contact with wheel with one hand while removing nuts with the other hand. 2. Step back when removing wheel so feet are out of the way.	Gloves
Fit replacement wheel and insert wheel nuts, hand-tightened	1. Injury from finger jammed in wheel nut/hole interface	1. Keep fingers back on end of wheel nut, away from threads.	Gloves
Lower jack and remove from vehicle	1. Injury on hand jammed between jack and vehicle base 2. Injury due to incorrectly aligned or inserted jack	1. Check that jack is securely positioned on vehicle base and on a flat surface before lowering. 2. Keep hands on jack handle or away from vehicle base while lowering.	Gloves
Tighten wheel nuts using correct torque measurement	1. Injury from finger jammed in wheel nut/hole interface	1. Keep fingers back on end of wheel nut, away from threads.	Gloves
Put damaged wheel and tools into storage space	Injury from lifting heavy objects.	1. Roll damaged wheel to storage (trunk); orient damaged wheel in upright position; stand as close as possible to wheel; bend knees and lift wheel carefully and slowly.	Gloves

JHA by: _____

Date: _____

Figure 6-2. Example Job Hazard Analysis Form for Changing a Tire

Table 6-2. OSHA Hazard Communication Standard (HCS) Details for SDSs

Section	Title	Contains
1	Identification	Product identifier; manufacturer or distributor name, address, telephone number; emergency telephone number; recommended use; restrictions on use.
2	Hazard(s) identification	All hazards regarding the chemical; required label elements.
3	Composition/ information on ingredients	Information on chemical ingredients; trade secret claims.
4	First-aid measures	Important symptoms/effects, acute, delayed; required treatment.
5	Fire-fighting measures	Suitable extinguishing techniques, equipment; chemical hazards from fire.
6	Accidental release measures	Emergency procedures; protective equipment; proper methods of containment and cleanup.
7	Handling and storage	Precautions for safe handling and storage, including incompatibilities.
8	Exposure controls/ personal protection	OSHA's permissible exposure limits (PELs); American Conference of Governmental Industrial Hygienists (ACGIH) Threshold Limit Values (TLVs); and any other exposure limit used or recommended by the chemical manufacturer, importer, or employer preparing the SDS, where available, as well as appropriate engineering controls; personal protective equipment (PPE).
9	Physical and chemical properties	The chemical's characteristics.
10	Stability and reactivity	Chemical stability and possibility of hazardous reactions.
11	Toxicological information	Routes of exposure; related symptoms, acute and chronic effects; numerical measures of toxicity.
12	Ecological information	Details regulated by particular agencies.
13	Disposal considerations	Details regulated by particular agencies.
14	Transport information	Details regulated by particular agencies.
15	Regulatory information	Details regulated by particular agencies.
16	Other information	Such as the date of preparation or last revision.

Hazardous Materials Identification System

The Hazardous Materials Identification System (HMIS) is a numerical hazard rating to label hazardous materials.

An HMIS sign shows four color-coded bars representing different hazard areas:

- **Blue** – Health
- **Red** – Flammability
- **Orange** – Physical hazard
- **White** – Personal protection

For health, flammability, and physical hazards, the sign shows a rating number from 0 (lowest risk) to 4 (highest risk). For personal protection, a letter is shown to indicate what type of PPE is to be used.

Figure 6-3 shows a typical HMIS sign. (Note that the colors defined above are not shown in this black and white image.) The chemical name is shown at the top of the sign and the risk numbers are shown alongside each hazard area. Table 6-3 shows the definition of risk ratings for each hazard area.

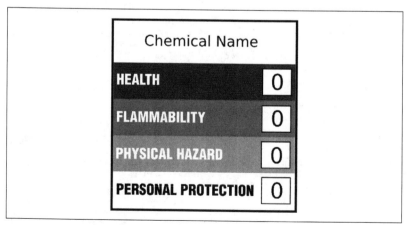

Figure 6-3. Blank HMIS Sign

Table 6-3. Risk Rating Definitions for Each HMIS Hazard Area

Color/Hazard Area	Associated Definitions
Blue/Health	4 – Life-threatening, major, or permanent damage may result from single or repeated overexposures (e.g., hydrogen cyanide). 3 – Major injury likely unless prompt action is taken and medical treatment is given. 2 – Temporary or minor injury may occur. 1 – Irritation or minor reversible injury possible. 0 – No significant risk to health. * – An asterisk alongside the number indicates that there is a chronic health hazard (e.g., risk of emphysema).
Red/Flammability	4 – Flammable gases, or very volatile flammable liquids with flash points below 73°F (23°C), and boiling points below 100°F (38°C). Materials may ignite spontaneously with air (e.g., propane). 3 – Materials capable of ignition under almost all normal temperature conditions. Includes flammable liquids with flash points below 73°F (23°C) and boiling points above 100°F (38°C), as well as liquids with flash points between 73°F and 100°F. 2 – Materials that must be moderately heated or exposed to high ambient temperatures before ignition will occur. Includes liquids having a flash point at or above 100°F (38°C) but below 200°F (93°C) (e.g., diesel fuel). 1 – Materials that must be preheated before ignition will occur. Includes liquids, solids, and semi-solids having a flash point above 200°F (93°C) (e.g., canola oil). 0 – Materials that will not burn (e.g., water).
Orange/Physical Hazard	4 – Materials that are readily capable of explosive water reaction, detonation or explosive decomposition, polymerization, or self-reaction at normal temperature and pressure. 3 – Materials that may form explosive mixtures with water and are capable of detonation or explosive reaction in the presence of a strong initiating source. Materials may polymerize, decompose, self-react, or undergo other chemical change at normal temperature and pressure with a moderate risk of explosion. 2 – Materials that are unstable and may undergo violent chemical changes at normal temperature and pressure with low risk for explosion. Materials may react violently with water or form peroxides upon exposure to air. 1 – Materials that are normally stable but can become unstable (self-react) at high temperatures and pressures. Materials may react nonviolently with water or undergo hazardous polymerization in the absence of inhibitors. 0 – Materials that are normally stable, even under fire conditions, and will not react with water, polymerize, decompose, condense, or self-react. Nonexplosives.

Table 6-3. Risk Rating Definitions for Each HMIS Hazard Area (*Continued*)

Color/Hazard Area	Associated Definitions
White/ Personal Protection	A – Safety glasses B – Safety glasses and gloves C – Safety glasses, gloves, and an apron D – Face shield, gloves, and an apron E – Safety glasses, gloves, and a dust respirator F – Safety glasses, gloves, apron, and a dust respirator G – Safety glasses and a vapor respirator H – Splash goggles, gloves, apron, and a vapor respirator I – Safety glasses, gloves, and a dust/vapor respirator J – Splash goggles, gloves, apron, and a dust/vapor respirator K – Airline hood or mask, gloves, full suit, and boots X – Ask supervisor or safety specialist for handling instructions

Process Safety

Hazardous Process Controls

Many mission critical operations involve multiple safety hazards, such as:

- Exposure to hazardous chemicals
- Exposure to high-pressure and high-temperature conditions
- Dangers relating to fire or explosion

Hazardous process controls are a critical part of the process safety management system in mission critical organizations. The entire process safety management system will incorporate many different types of safeguards:

- Basic process control system (BPCS), to maintain key process variables in their normal operating envelope
- Operator supervision and intervention, to respond to alarm conditions reported by the BPCS

- Safety instrumented system (SIS), to automatically intervene if manual efforts fail to control dangerous conditions

- Independent fail-safe mechanisms, such as relief valves, that will operate in the event that the SIS intervention does not prevent the dangerous condition from persisting

- Containment systems, such as dikes or bunkers, that can hold any materials released during an incident, to prevent or minimize the impact to the facility or wider community

- Facility and community response procedures that will be initiated in emergency situations

This is represented in Figure 6-4.

Fail-Safe Mechanisms

Table 6-4 lists the common fail-safe mechanisms used as part of an overall process safety management system.

Process Hazard Analysis

A process hazard analysis (PHA) is a structured assessment of the potential hazards associated with an industrial process. The PHA process is sometimes called a hazard and operability study (HAZOP).

The PHA process analyzes potential causes and consequences of all potential hazards, considering all aspects—equipment, instrumentation, operator error, and any external factors (e.g., weather)—that might impact the process. The PHA process now considers cybersecurity-related factors, as

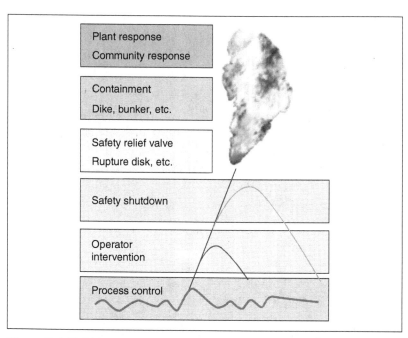

Figure 6-4. Relationship Between Various Hazardous Process Controls

Table 6-4. Common Fail-Safe Mechanisms

Mechanism	Description
Interlock	A feature that makes the state of two mechanisms or functions mutually dependent. It may be used to prevent undesired states in a finite-state machine, and may consist of any electrical, electronic, or mechanical devices or systems. In most applications, an interlock is used to help prevent a machine from harming its operator or damaging itself by preventing one element from changing state due to the state of another element, and vice versa.
Pressure relief valve (PRV)	A type of valve used to control or limit the pressure in a system that can build up for a process upset, instrument or equipment failure, or fire. Allowing the pressurized material to flow from an auxiliary passage out of the system relieves the pressure. The relief valve is designed to open at a predetermined pressure to protect vessels or other equipment from being subjected to pressures that exceed their design limits.
Rupture disk	This is a particular type of pressure relief device that protects a vessel or system from over- or under-pressurization. A rupture disk is a one-time-use device, as it fails when subjected to a predetermined pressure.

these can present a significant threat to modern mission critical systems.

PHA processes will vary from organization to organization and they may use a variety of methods to obtain a result, depending on the process. Common methods include:

- **Failure Modes and Effects Analysis (FMEA)** – This involves reviewing as many system elements as possible to identify their failure modes, and their causes and effects. For each component, the failure modes and their resulting effects on the rest of the system are recorded in a form.

- **Layer of Protection Analysis (LOPA)** – This method accounts for each identified hazard by documenting the initiating cause and the protection layers that prevent or mitigate the hazard. The total amount of risk reduction can then be determined and the need for more risk reduction analyzed.

- **Fault Tree Analysis (FTA)** – This method analyzes undesired states of a system using Boolean logic to combine a series of lower-level events. For example, if a failure condition requires that two pumps fail simultaneously, the logic would be PUMP1 AND PUMP2; whereas if the failure requires only one pump to fail, it would be PUMP1 OR PUMP2.

In the United States, OSHA mandates the PHA process in its process safety management (PSM) regulation for the identification of risks involved in the design, operation, and modification of processes that handle highly hazardous chemicals. The PHA process is discussed in more depth in Chapter 7, "Fundamentals of Risk Management."

Safety Instrumented Functions and Safety Integrity Levels

Mission critical systems that contain safety-related hazards require dedicated equipment to reduce the risk due to specific hazards. Examples include:

- Vent valve and associated monitoring, and control equipment that ensures the valve operates in the event of a high-pressure condition

- Fuel trip and associated monitoring, and control equipment that ensures the trip operates in the event of a high-temperature condition

Each group of equipment that is deployed to manage a particular hazard is called a safety instrumented function (SIF).

The definition of the level of risk reduction a SIF provides, or is required to provide, is called the safety integrity level (SIL). SIL may be represented in various ways, such as:

- The required availability of the SIF.

- The probability of failure of the SIF (known as the probability of failure on demand—PFD).

- The level of risk reduction that the SIF provides (known as the risk reduction factor—RRF).

- Qualitative measures, such as the number of fatalities that could occur if the SIF fails. Such measures will be dependent on the organization, location, process, and other specific factors.

The International Electrotechnical Commission IEC 61508 standard defines four SIL ratings, from 1 to 4, where SIL 1 is the least dependable and SIL 4 is the most dependable. The

International Society of Automation ISA84 standards define SIL 1 through SIL 3 only. Table 6-5 shows the IEC SIL ratings and associated definitions, including example qualitative measures.

A SIL is assigned after performing a risk analysis to determine the risk associated with a specific hazard that the SIF is to protect against.

Initially, the risk is calculated without the risk reduction effect of the SIF. The resulting risk is compared against a risk target. The difference between the calculated risk and the risk target must be addressed through risk reduction of the SIF.

Hazardous Area Classification

A hazardous area is defined as an area where there is a potential for an explosive or flammable atmosphere to exist. Such atmospheres may exist normally (e.g., coal mines) or under fault conditions (e.g., in a petrochemical refinery where there is a leak in a tank holding flammable gas or liquid).

Equipment that is to be installed in hazardous areas must be purposely designed for use within those areas. For example,

Table 6-5. Safety Integrity Level Definitions

SIL	Availability	Probability of Failure on Demand (PFD)	Risk Reduction Factor (RRF)	Qualitative Measure
1	90%–99%	10^{-1} to 10^{-2}	10–100	Potential for minor onsite injuries
2	99%–99.9%	10^{-2} to 10^{-3}	100–1000	Potential for major onsite injuries
3	99.9%–99.99%	10^{-3} to 10^{-4}	1000–10,000	Potential for multiple onsite fatalities
4	>99.99%	10^{-4} to 10^{-5}	10,000–100,000	Potential for community fatalities

electrical or electronic equipment must be designed such that it cannot generate sufficient energy to ignite the explosive or flammable atmosphere. The classification of the area, and thus the requirements for the equipment, vary. Hazardous area classification provides a means to define the area and its requirements.

The National Electrical Code (NEC) defines hazardous area classifications in the United States (NEC Article 500). An NEC hazardous area classification consists of several parts: the class, group, and division. Worldwide, outside the United States, IEC 60079 defines hazardous area classifications using class and zone (this classification method is known as ATEX, an abbreviation of the French term "atmosphères explosibles"). NEC has adopted the zone method as an alternative method of classification (NEC Article 505).

A typical U.S. classification example, using both NEC methods, is shown in Figure 6-5.

There are three classes:

- **Class I** – Locations in which flammable gases, flammable liquid vapors, or combustible liquid vapors may be

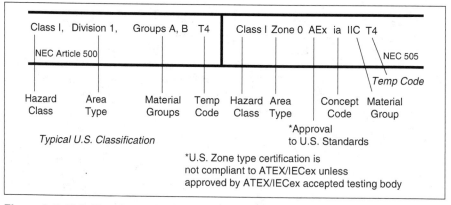

Figure 6-5. U.S. Hazardous Area Classification Example

present in quantities sufficient to produce explosive or ignitable mixtures

- **Class II** – Locations that are hazardous because of the presence of combustible dust
- **Class III** – Locations that are hazardous because of the presence of easily ignitable fibers or materials producing combustible fibers

For these three classes, there are two divisions:

- **Division 1** – Locations where hazardous material may exist under normal operating conditions, under frequent repair or maintenance operations, or where faulty operation of equipment might cause simultaneous failure of electrical equipment
- **Division 2** – Locations where hazardous material exists but is confined within a closed container or closed system from which it can only escape under accidental rupture, breakdown, or in abnormal operation

For class I, the following location groups are defined, based on the type of gas or vapor present:

- **Group A** – Acetylene
- **Group B** – Flammable gas, flammable liquid vapors, or combustible liquid vapors, such as hydrogen, ethylene oxide, and propylene oxide
- **Group C** – Flammable gas, flammable liquid vapors, or combustible liquid vapors, such as ethyl ether, ethylene, and acetaldehyde
- **Group D** – Flammable gas, flammable liquid vapors, or combustible liquid vapors, such as acetone, ammonia, and benzene

For class II and class III, the following location groups are defined, based on the type of dust or particle present:

- **Group E** – Metallic dusts, including aluminum, magnesium, and their commercial alloys, or other combustible dusts whose particle size, abrasiveness, and conductivity present similar hazards in the use of electrical equipment

- **Group F** – Carbonaceous dusts that have more than 8% total entrapped volatiles (e.g., coal, charcoal, and coke dust)

- **Group G** – Combustible dusts not included in Group E or F (e.g., flour, grain, wood, plastic, and chemicals)

The zone classifications are defined as shown in Table 6-6.

Temperature is a significant factor in hazardous area classification. The temperature code (T4 in the example in Figure 6-5)

Table 6-6. Hazardous Area Zone Definitions

Zone	Material	Zone Definition
0	Flammable gases, flammable liquid vapors, or combustible liquid vapors	The explosive atmosphere is present continuously or for long periods or frequently.
1		The explosive atmosphere is likely to occur in normal operation occasionally.
2		The explosive atmosphere is not likely to occur in normal operation but, if it does occur, it will persist for a short period only.
20	Combustible dust	The explosive atmosphere is present continuously or for long periods or frequently.
21		The explosive atmosphere is likely to occur in normal operation occasionally.
22		The explosive atmosphere is not likely to occur in normal operation but, if it does occur, it will persist for a short period only.

Table 6-7. Hazardous Area Temperature Code Definitions

Class	Maximum Surface Temperature	
	(°C)	(°F)
T1	450	842
T2	300	572
T3	200	392
T4	135	275
T5	100	212
T6	85	185

defines the maximum allowable surface temperature for the equipment, as shown in Table 6-7.

There are several approved methods to prevent equipment in hazardous areas from causing an ignition of the explosive or flammable material. In the international ATEX classification system, these protection methods are assigned a unique code, which is shown on the equipment label. Table 6-8 provides a list of these methods and their codes.

Environmental Safety

Emissions and Discharges

Many mission critical organizations operate processes that present hazards to the environment, such as:

- Pollution of rivers and water sources
- Pollution of the atmosphere

Those organizations are required to follow all relevant regulations to control emissions and discharges to acceptable and safe levels.

Table 6-8. Hazardous Area Equipment Protection Methods and Associated ATEX Codes

Method	Description	ATEX Code
Encapsulated	The components are encased in a resin-type material.	Ex m
Flameproof	The components are placed in an enclosure that can withstand the pressure developed during an internal explosion of an explosive mixture. The explosion is not transmitted to the explosive surrounding atmosphere, and the enclosure operates with a temperature too low for the surrounding explosive gas or vapor to ignite.	Ex d
Intrinsically safe	Electrical equipment is designed so that, under normal or abnormal conditions, it is incapable of releasing sufficient electrical or thermal energy to cause an ignition of the hazardous surrounding atmospheric mixture.	Ex i
Increased safety	Applies only to high-quality and very robust components. Various measures are applied to reduce the probability of excessive temperatures and the occurrence of arcs or sparks in the interior and on the external parts of the equipment. Increased safety may be used with flameproof type of protection.	Ex e
Oil-filled	Protection is applied by submerging the components in oil.	Ex o
Pressurized/purged	Protection is applied by maintaining a positive pressure in the enclosure with air or an inert gas, preventing the surrounding ignitable atmosphere from coming into contact with energized parts of the equipment.	Ex p
Sand/powder/quartz filled	Protection is applied by covering the components with sand, powder, or quartz.	Ex q
Nonincendive	A protection applied to electrical equipment in such a way that normal operation is not capable of igniting surrounding explosive atmospheres.	Ex n
Special protection	Any method that can be shown to have the required degree of safety.	Ex s

Continuous emission monitoring systems (CEMSs) are common in organizations where air emission standards compliance is required; for example, monitoring carbon monoxide and carbon dioxide, sulfur dioxide, nitrogen oxide, hydrogen chloride, airborne particulate matter, mercury, volatile organic compounds, and oxygen.

Organizations that discharge to surface waters are required to produce a periodic discharge monitoring report (DMR). This involves collecting wastewater samples and conducting chemical and biological tests of these samples. In some organizations, continuous monitoring systems may be in place to collect such data in real time.

Loss of Containment

A loss of containment is an unplanned or uncontrolled release of material from primary containment, including nontoxic and nonflammable materials (e.g., steam, hot condensate, nitrogen, compressed carbon dioxide, or compressed air). Loss of containment of hazardous materials is the most common cause of process incidents.

Loss of containment incidents occur because of:

- Mechanical integrity failures
- Poor process design
- Incomplete process technology documentation
- Inadequate hazard analysis
- Lack of management of change
- Unexpected or uncontrolled reactions
- Human error

The multiple layers of protection (BPCS, operator monitoring, SIS, independent fail-safes, containment systems, and emergency response) are designed to mitigate these risks.

Safe Handling and Disposal of Materials

Many mission critical organizations utilize materials that must be carefully handled and disposed of when it is no longer required or expired. Typical materials include:

- Hazardous chemicals and their containers
- Radioactive materials
- Batteries, especially lithium-based cells
- Fluorescent light bulbs and tubes
- Paint
- Medical sharps
- Waste oil and oil filters

Organizations should have defined procedures that detail:

- How and where materials are stored, including the material required for the container
- Labeling for the materials, including expiry date (if applicable)
- Safety procedures in the event of an incident with the material
- Disposal procedures

Public Safety

Mission critical organizations operate within communities. Public safety is of paramount concern where hazards exist that can impact the community.

The U.S. government has structures in place to deal with incidents, and mission critical organizations will have a response plan that takes this into account, when dealing with state, tribal, and local authorities.

Typical public safety activities for mission critical organizations to consider include:

- **Pre-incident coordination** – Supporting incident management planning activities
- **Emergency response** – Including medical, fire, and rescue services
- **Site security** – Providing security forces and establishing protective measures around the incident site, critical infrastructure, and/or critical facilities
- **Traffic and crowd control** – Providing emergency protective services to address public safety and security requirements
- **Specialized resources** – Providing specialized assets, such as traffic barriers and chemical, biological, radiological, nuclear, and explosive detection devices

Physical Security

Physical security is the protection of personnel, equipment, and data from physical actions and events that could cause serious loss or damage to an organization. This includes protection from burglary, theft, vandalism, and terrorism (foreign and domestic).

Physical security has three key elements:

1. Access control
2. Intrusion detection
3. Incident response

Access Control

Organizations must secure themselves from potential threats using a variety of measures, such as:

- Perimeter fencing
- Locks on doors, gates, and equipment cabinets
- Access control systems, using key cards and biometric methods

Multiple levels of security should be used where possible to reduce the likelihood of a breach in certain areas, such as control rooms or critical equipment cabinets.

To minimize the risks of unauthorized access, organizations should limit area access to only those who need it to perform their duties.

Intrusion Detection and Prevention

Physical locations should be continuously monitored using:

- Camera systems
- Access notification systems
- Motion detection and heat sensors

Electronic logs and records should be maintained and reviewed regularly. Although actual security breaches may not have occurred, unusual or prohibited behavior may be observed from these records and could be used to prevent a future breach.

Incident Response

Organizations must have procedures in place to respond to physical security incidents. These procedures may involve collaboration with external authorities, and appropriate contact information must be maintained to ensure it is available when required.

These procedures should be tested on a regular basis to ensure they are effective and the necessary information is available should it be required in the event of a real incident.

Review Questions

6.1 What is the collective term for equipment worn by personnel to protect against workplace hazards?

  A. Personal protective equipment (PPE)

  B. Personnel protective equipment (PPE)

  C. Personal protection equipment (PPE)

  D. Practical protection equipment (PPE)

6.2 On a lockout/tagout (LOTO) tag, what color is used to signify biological hazards?

  A. Red

  B. Yellow

  C. Orange

  D. Fluorescent orange

6.3 What is the term for the process used to identify potential hazards in an activity?

  A. Personnel hazard analysis (PHA)

  B. Activity hazard analysis (AHA)

  C. Job safety analysis (JSA)

  D. Process hazard analysis (PHA)

6.4 What is the name of the standard that defines the details that must be provided on Safety Data Sheets (SDSs)?

A. Safety Data Standard (SDS)

B. Hazard Communication Standard (HCS)

C. Material Safety Standard (MSS)

D. Chemical Safety Standard (CSS)

6.5 On a Hazardous Materials Identification System (HMIS) label, what color indicates a physical hazard?

A. Blue

B. Red

C. Orange

D. White

6.6 In process safety, what is the name for the safety feature that makes the state of two mechanisms or functions mutually dependent?

A. Interlock

B. Control

C. Feedback

D. Feedforward

6.7 In process safety, what is the name of the equipment that is used to manage a particular hazard?

A. Safety integrity level (SIL)

B. Safety instrumented system (SIS)

C. Emergency shutdown (ESD)

D. Safety instrumented function (SIF)

6.8 In IEC 61508, what is the required availability of a safety integrity level (SIL) 4 SIF?

A. 90%–99%

B. 99%–99.9%

C. 99.9%–99.99%

D. >99.99%

6.9 In U.S. hazardous area classification, what group covers metallic dusts?

A. Group C

B. Group D

C. Group E

D. Group F

6.10 For hazardous area equipment, what is the method that ensures that electrical equipment is incapable of releasing sufficient energy to cause an ignition under normal or abnormal conditions?

A. Intrinsically safe

B. Increased safety

C. Nonincendive

D. Special protection

7
Fundamentals of Risk Management

Definition of Risk

In a mission critical environment, the typical risks include:

- Death or injury
- Equipment damage
- Environmental violations
- Business interruption

Risk is "the effect of uncertainty on objectives." Risk management is "the coordinated set of activities and methods that is used to direct an organization and to control the many risks that can affect its ability to achieve objectives."

Consider the definition of risk in the context of safety and security:

- **Safety definition** – Risk is a measure of human injury, environmental damage, or economic loss in terms of

Table 7-1. Risk Management Resources

Federal Emergency Management Agency (FEMA)	Risk Management Series
International Organization for Standardization (ISO)	ISO 31000 – *Risk Management*
National Institute of Standards and Technology (NIST)	NIST SP 800-37 – *Managing Information Security Risk*
National Institute of Standards and Technology (NIST)	NIST SP 800-30 – *Guide for Conducting Risk Assessments*
Department of Homeland Security (DHS)	Risk Management Fundamentals

both the incident likelihood and the magnitude of the loss or injury.

- **Security definition** – Risk is an expression of the likelihood that a defined threat will exploit a specific vulnerability of a particular attractive target or combination of targets to cause a given set of consequences.

Table 7-1 provides key resources available to help manage risk within an organization.

The Risk Management Cycle

Risk management is a continuous process, described by the risk management cycle below. Risks are identified and then assessed to determine their severity to the organization. As a result of the assessment, controls are identified to manage these risks. In order to effectively manage risk, the risks and the associated controls must be reviewed on a regular basis to determine:

- If any risk has occurred
- If each risk is still present
- If the severity of each risk has changed

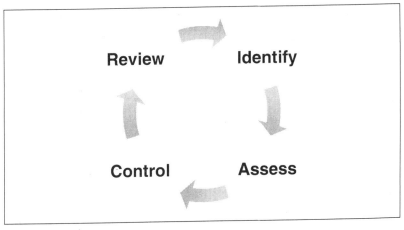

Figure 7-1. The Risk Management Cycle

- If the controls are in place
- If the controls are effective

Based on this review, controls may be changed or added. The cycle (see Figure 7-1) continues as new risks are identified and managed.

Risk Management Components

Table 7-2 lists the key components of risk management plans.

In most corporations, risk management systems will already have developed scales for likelihood and consequence. Table 7-3 shows what a typical likelihood scale might be for an organization.

In this example, likelihood is expressed in terms of time frequency. It may also be expressed as a probability (e.g., 1 in 100) or percentage (e.g., 10%). The main objective of the likelihood scale is to convert whatever the measure is into a consistent scheme that is easy to understand.

Table 7-2. Key Components of Risk Management Plans

Risk	Risk is the potential for something bad to occur. Risk is the combination of two factors: the likelihood that an event will occur and the consequence of that event occurring. Risk is often described in terms of an equation, Risk = Likelihood • Consequence.
Likelihood	Likelihood is the estimate or probability that an event will occur. This may be written as a probability (1 in 100), a percentage (10%), or in some other meaningful way (e.g., "once in 10 years").
Consequence	Consequence is the expected impact of an event occurring. This may be expressed as a financial value, an impact on workers (e.g., number injured or killed), or in some other way that is meaningful to the organization concerned.
Hazard	A hazard is a source of potential damage or harm. Hazards can be natural (e.g., a hurricane or flood) or man-made (e.g., an exposed electrical circuit).
Threat	A threat is another name for a hazard, but it is particularly common in the management of cybersecurity risk. Threats can be accidental (a user making a mistake) or deliberate, and can be caused by insiders (employees) or outsiders (hackers). Examples of cybersecurity threats include: • Malware • Information disclosure • Denial of service • Unauthorized access
Vulnerability	A vulnerability is a flaw in a device, in software, or in a process that affects how likely it is that a hazard or threat can cause an event. For example, a process that does not require a technician to turn off the supply to equipment before performing maintenance can result in the technician being exposed to the hazard of an exposed electrical circuit, resulting in electrocution. In cybersecurity, a vulnerability may be a bug in software that malware can exploit to allow unauthorized access to a system.
Mitigation	Mitigation is the process of reducing the likelihood or consequence of an event. For example: • The introduction of redundant equipment will reduce the likelihood of a failure of that equipment. • The introduction of an independent fail-safe mechanism (such as a pressure relief valve) will reduce the consequence of a failure of primary equipment.

Table 7-2. Key Components of Risk Management Plans (*Continued*)

Risk Tolerance	The level of risk the organization is willing to tolerate. The combination of likelihood and consequence will be used to determine risk tolerance. For example, one organization may decide that it is willing to tolerate a risk that may cost $500,000 and may occur once a year, whereas another organization may decide that it is only willing to tolerate such a cost once every 10 years.
Risk Response	Risks can be addressed in several ways. An organization's risk response can be: • To design the risk out, for example by replacing components that contribute to the risk. • To reduce the risk, for example, by introducing additional measures such as equipment redundancy. • Accept the risk, noting that it may occur but monitoring it. • Transfer or share the risk, for example, management of specific equipment could be delegated to a third party with particular capabilities, or insurance could be purchased to cover the costs of an event occurring.

Table 7-4 shows an example consequence scale for an organization.

For most organizations, the result of a risk occurring may have multiple consequences. For instance, there may be an impact on workers (who could be injured or killed) or the environment (e.g., due to an oil spill) or legal implications (e.g., regulatory fines, criminal proceedings). As a result, a typical

Table 7-3. Typical Likelihood Scale for an Organization

Category	Description
High	A threat/vulnerability whose occurrence is probable in the next year
Medium	A threat/vulnerability whose occurrence is probable in the next 10 years
Low	A threat/vulnerability whose occurrence is probable in the next 100 years
Not Applicable	A threat/vulnerability for which there is no history of occurrence and for which the probability of occurrence is deemed extremely unlikely

Table 7-4. Example Consequence Scale for an Organization

Category	Business Continuity Planning		Information Security			Process Safety		Environmental Safety
	Outage – Single Site	Outage – Multiple Sites	Cost	Legal	Public Confidence	People – On-site	People – Off-site	Environment
High	>7 days	>1 day	>$500 million	Criminal offense – felony	Loss of brand image	Fatality	Fatality or major community incident	Citation by regional or national agency or long-term damage over large area
Medium	>2 days	>1 hour	>$5 million	Criminal offense – misdemeanor	Loss of customer confidence	Lost workday or major injury	Complaints or local community impact	Citation by local agency
Low	<1 day	<1 hour	<$5 million	None	None	First aid or recordable injury	No complaints	Small, contained release below reportable limits

consequence scale will show the relationship of the simple, consistent scheme to actual impacts in various categories.

The specific values will vary from organization to organization. For example, a large organization may consider a high impact to be a cost of more than $500 million whereas a small organization may decide that high impact is a cost of more than $500,000.

The categories may also vary from organization to organization. For example, some organizations may identify consequences related to stockholders or regulators.

Having determined what high, medium, and low likelihood and consequence mean, the organization must then decide on the appropriate level of response to risk, the combination of these two values. This is typically done using a matrix such as that shown in Table 7-5, which considers all possible combinations of likelihood and consequence value in a three-by-three matrix. Some organizations may choose to represent their risks with a larger matrix (e.g., five-by-five, eight-by-eight).

The organization may decide that:

- Risks categorized as A and B are intolerable and therefore must be addressed through one or more risk mitigations so that the revised categorization is reduced to the C or D level.

Table 7-5. Typical Risk Matrix for an Organization

		Consequence		
		High	Medium	Low
Likelihood	High	A	B	C
	Medium	B	C	D
	Low	C	D	D

- Risks categorized as C may be tolerable and therefore do not require addressing but will need to be monitored carefully.

- Risks categorized as D are sufficiently low that they do not need special attention.

Quantitative and Qualitative Risk Analysis

Qualitative risk analysis relies on the input of experienced employees and/or experts to provide information regarding likelihood and severity of specific threats impacting specific assets. Different levels of likelihood and severity are identified using a general classification, such as high, medium, and low, rather than specific estimates.

Quantitative risk analysis relies on extensive data sets that document the rate at which damage occurs to assets based on exposure to defined combinations of threats and vulnerabilities. This method can provide more precise risk estimates than qualitative risk analysis methods but requires more data to be available and is only as good as that data.

Process Hazard Analysis

The process hazard analysis (PHA) is a risk assessment process used in many mission critical organizations, particularly those involved in industrial processes such as water, wastewater, and oil and gas processing. PHA is traditionally associated with assessing the risk of safety-related incidents, such as chemical spills, fires, or explosions.

As with other types of risk assessment, the objective of a PHA is to identify potential threats or hazards and to assess the likelihood and consequence of various events occurring (see Figure 7-2).

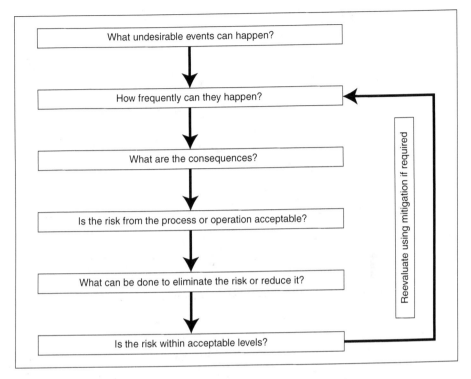

Figure 7-2. Example Process Hazard Analysis (PHA) Flowchart

Further details of PHA are provided in Chapter 6, "Safety and Physical Safety."

Identification of Mission Critical Assets

Identification of assets is a critical stage in the assessment of risk in mission critical organizations. It is essential that organizations understand:

- What systems they have and what components make up these systems
- How the systems are connected and how users access them
- The impact of system failure for each system

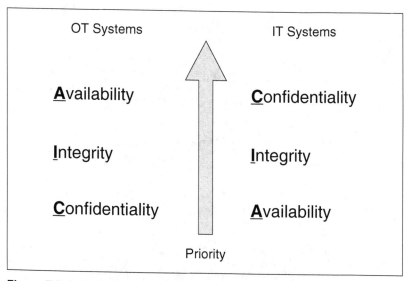

Figure 7-3. Relative Priorities for IT and OT Systems

The availability, integrity, and confidentiality factors of a system are key to determining risk. As a reminder from Chapter 2, "Mission Critical Operations Concepts," the relative importance of these factors varies for operational technology (OT) and information technology (IT) systems, as shown in Figure 7-3.

Hazard and Threat Assessment

To help identify and assess potential hazards or threats, it is useful to categorize them. In general, at the highest level, hazards can be internal and external to the organization. Within these high-level categories, hazards can be further subdivided until there is a clear set of groups that can then be used by the organization to identify specific hazards that affect them. An example of this grouping is shown in Figure 7-4, from the ANSI/ISA-62443 international standards, developed by the International Society of Automation (ISA).

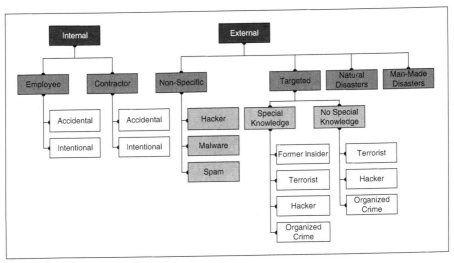

Figure 7-4. The ANSI/ISA-62443 Taxonomy of Threats

Vulnerability Assessment

Having identified the systems and associated devices, the organization must assess the vulnerabilities in these systems and the associated processes. Vulnerability assessment must include:

- Systems vulnerability analysis, considering the component equipment (e.g., what operating systems are used and what known issues exist with these?) and the architecture (e.g., are there any single points of failure or devices that, if failed, can cause a system outage?)
- Physical vulnerability analysis, considering the environment (e.g., is the location within an earthquake or tornado zone?)
- Process vulnerabilities, including such aspects as management of portable media, management of usernames and passwords, and change control

Much of this work can be conducted as a desktop exercise using vendor documentation and procedure documents. For a thorough evaluation of vulnerabilities, penetration testing is recommended. Penetration testing involves the deliberate attempt, usually by specialist external third parties, to identify and exploit vulnerabilities to understand the level of exposure an organization has.

Risk Assessment

The objective of risk analysis is to:

- Determine which assets must be protected (people, processes, equipment, information, chemicals, etc.)
- Determine the consequence of a compromise for each of the assets (loss of production, health/safety impact, environmental impact, etc.)
- Determine the vulnerability of those assets, taking into account existing safeguards
- Determine the threats to those assets (theft, misuse, damage, system malfunction, etc.)
- Recommend changes that reduce risk to an acceptable level
- Determine priorities for activities to manage risk
- Provide a foundation for building a security policy and plan

Risk Management Plans

Having identified the likely consequences of risks, it is necessary to identify sufficient mitigations or countermeasures to address these risks. Installing more technology is not the only option; the organization can:

- Mitigate the risk through new processes or procedures
- Avoid the risk by eliminating the root cause and/or consequence
- Transfer the risk (i.e., use other options to compensate for the possible loss, such as purchasing insurance)

As with safety, it is impossible to avoid risk in mission critical systems operation. The objective of risk reduction is to identify the balance between the cost of implementing and maintaining measures to mitigate the risk against the potential impact to the business.

In risk management, the term *as low as reasonably practicable* (ALARP) is used to define the point at which it is not possible to realistically reduce risk any further (see Figure 7-5).

For example, while it is obviously necessary to spend $1 million to prevent a major explosion capable of killing 150 people, spending the same amount to prevent an equipment failure that is expected to occur once every 100 years will not be.

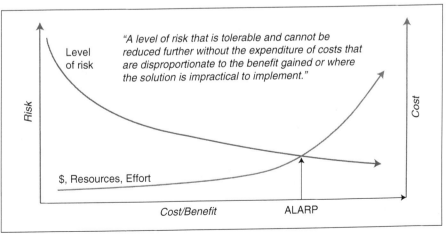

Figure 7-5. Visualizing the ALARP Assessment of Risks

Review Questions

7.1 What are the steps in the risk management life cycle?

    A. Identify, assess, control, and review

    B. Identify, protect, detect, and respond

    C. Identify, measure, monitor, and review

    D. Identify, measure, detect, and respond

7.2 What are the key elements of the risk equation?

    A. Likelihood and vulnerability

    B. Hazard and threat

    C. Hazard and consequence

    D. Likelihood and consequence

7.3 What is the activity called that analyzes the potential issues in a system?

    A. Risk assessment

    B. Threat assessment

    C. Vulnerability assessment

    D. Process assessment

7.4 What is a form of risk response that involves purchasing insurance?

    A. Design

    B. Reduce

    C. Accept

    D. Transfer

7.5 What is the term that defines the point at which it is not possible to realistically reduce risk any further?

A. As low as reasonably practicable (ALARP)

B. As low as realistically possible (ALARP)

C. As limited as reasonably practicable (ALARP)

D. As limited as realistically possible (ALARP)

8
Continuity of Operations and Emergency Response

Business Resilience Planning

The Purpose of Business Resilience Planning

Emergency response begins with planning to prepare for mitigating the impact of events that cause a disruption to operations. This effort may be known as *business resiliency*, continuity of operations (COO), business continuity (BC), or *service continuity planning*. Planning ensures that an organization is ready to anticipate, respond to, and recover from an incident. The types of incidents that are likely—as well as the resulting response and recovery activities—will vary from organization to organization.

Some typical incident categories that critical infrastructure organizations will need to consider are:

- Man-made disasters, such as earthquakes, floods, or hurricanes

- Terrorist attacks

- Process incidents, such as loss of containment, fire, or explosion

- Loss of critical power or other resources, such as water supply

- Cybersecurity incidents, such as deliberate attacks or accidental events

Business resiliency requires an understanding of what the organization needs to continue its core functions. For example, while it might seem that a primary objective would be to restore the telephone service for a corporate office, it is more likely that this will be a lower priority than restoring that service to an operations center, without which the organization could not function. Business resilience plans extend beyond the organization, to include product and service suppliers and varying levels of government.

COO and emergency response planning consists of four key aspects:

1. **Business impact analysis** – The purpose of business impact analysis (BIA) is to identify the essential functions and resources in the organization, and to consider threats and vulnerabilities related to them. Getting this aspect correct, above all, is crucial. Omitting an essential function or resource, or neglecting a potential threat or vulnerability, will result in a lack of preparedness that could have serious consequences for a critical infrastructure organization.

2. **Business continuity planning** – The aim of a BC plan (also often referred to as a continuity of operations

plan—COOP) is to guarantee there is a process in place to ensure critical business functions can continue in the event of a serious incident, where disaster recovery (DR) plans will require long-term or major activities. A typical example of a BC plan is to establish critical business functions at a secondary location (e.g., in the event of major flood or storm damage to the primary location).

3. **Incident response** – The aim of an incident response (IR) plan is to identify the activities required during or immediately after an emergency. The focus at this time will be on activities such as containment, emergency treatment for affected individuals, and communications and alerts.

4. **Disaster recovery** – A disaster recovery (DR) plan takes over from an IR plan and is focused on restoration activities, such as reestablishing communications networks, IT equipment, or process operations.

Training around these emergency response-related plans (BC or COOP, IR, and DR) is crucial in a mission critical environment. Operations and management staff must be aware of the plans and activities required in the case of an emergency. In addition, the response activities should be tested to ensure that the staff knows the procedures and that the resources needed for response will be available and operational. These validation exercises could include efforts such as disaster recovery drills, testing of operations from alternative locations, and audits of offsite data storage.

The process of emergency response planning relies heavily on the principles of risk management (described in Chapter 7, "Fundamentals of Risk Management").

Identifying and Prioritizing Essential Functions and Resources

The purpose of a BIA is to identify an organization's essential functions and resources. This is necessary in order to understand:

- What, in the event of a major incident, will prevent a business from operating. For example, there may be some operational systems that cannot be operated manually, without an automation system.

- What the relative priorities are for the organization. For example, availability of the IT network is likely to be more critical than an asset related to human resources, such as a timesheet system.

The IR, DR, and BC plans will be based on this assessment; therefore, it is essential that this step is thorough and detailed. In order for this process to be effective, stakeholders from all business departments will need to provide input.

Although prioritizing essential functions and resources is very organization-specific, one option is to use a weighted scoring system, such as that shown in Table 8-1.

Table 8-1. Weighted Scoring to Indicate Relative Impact on Essential Functions

Function/ Resource	Contribution to/Impact on				
	Profitability	Strategic Objectives	Internal Operations	Reputation	Total
	40%	20%	10%	30%	100%
Function A	50	50	80	30	47
Function B	75	80	60	60	70
Function C	65	70	50	50	60
Resource A	30	30	80	30	35
Resource B	80	80	40	60	70

This method is somewhat subjective; however, it is the relative values that are important, rather than the absolute numbers. Provided all the relevant stakeholders are involved in the prioritization process, it should be possible to agree on such a table, which can then be used to identify the order of restoration for essential functions and resources.

Incident Response

Implementing an IR Plan

The IR plan will define:

- The scope of the plan, in terms of what organizational elements are included or excluded, and any circumstances or situations that are included or excluded
- A list of the incidents identified, together with their prioritization
- The organizational structure for the IR team, with a clear definition of roles, responsibilities, and levels of authority
- Communications procedures and contact information
- Reporting procedures and associated forms

The IR plan will identify the procedures for handling incidents and should categorize these procedures for three different stages:

1. **Before the incident occurs** – The IR plan identifies the activities that must take place to be prepared for an incident. An example of a pre-incident procedure is making regular backups of systems.

2. **While the incident is underway** – The IR plan identifies activities that must be performed in response to each

type of incident. An example of a procedure that occurs during the incident is verifying the details of malware on detection.

3. **Immediately after the incident** – The IR plan identifies activities that must take place before the incident can be defined as resolved, and forms the link between the IR plan and the DR plan, where recovery from the incident occurs.

Incident Command System

An emergency within a critical infrastructure organization may have severe implications for the wider community or environment. As a result, IR plans for critical infrastructure organizations will likely need to define the interaction with the National Incident Management System (NIMS) and the Incident Command System (ICS).

The NIMS was developed and is supported by the Department of Homeland Security (DHS) to enable the appropriate and timely response to large-scale incidents in the United States. The ICS is a key element of the NIMS.

The ICS is a management system designed to combine facilities, equipment, personnel, procedures, and communications from a wide range of governmental and nongovernmental organizations so that they can operate within a common organizational structure to enable effective and efficient domestic incident management.

The ICS structure (see Figure 8-1) defines a unified command consisting of key individuals, as well as individuals responsible for safety, operations, planning, logistics, finance, information, and liaison.

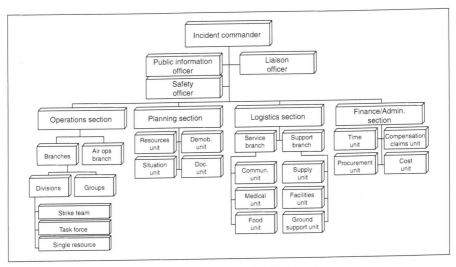

Figure 8-1. The Incident Command System (ICS) Structure

The Tiered Support Structure

The NIMS and ICS engage the support of the following groups:

- Federal government
- State government
- Local government
- Tribal government
- Private sector
- Nongovernmental organizations

Standards and regulations have been developed to ensure that all these groups can cooperate in the event of an emergency.

A key role in the IR team is the incident commander (IC). The IC should be an employee who is a member of management with the authority to make decisions. The IC is responsible for frontline management of the incident, as well as for determining whether outside assistance is needed and for relaying requests regarding internal resources or outside assistance.

Regulatory Compliance

Critical infrastructure organizations will usually have some form of regulatory compliance requirements relating to emergency management and incident response. Regulatory bodies define requirements for, among other things, reporting and communications. The regulatory body will depend on the industry; examples include:

- Environmental Protection Agency (EPA)
- Nuclear Regulatory Commission (NRC)
- North American Electric Reliability Corporation (NERC)
- Department of Homeland Security (DHS)

Stakeholder Communications

The IR plan will identify all stakeholders who will need to be involved or kept informed during an incident. Stakeholders may be:

- Employees
- Directors
- Shareholders
- Customers
- Suppliers
- Governmental and regulatory bodies

The methods of communication, the information communicated, and the frequency of communication may vary, depending on the stakeholder. The IR plan should define this level of detail and the stakeholders themselves should agree to it.

Disaster Recovery

Implementing a DR Plan

The DR plan, which goes into operation immediately after the IR plan is complete, focuses on activities for restoring normal operation. The DR plan will define:

- The scope of the plan, in terms of what organizational elements are included or excluded, and any circumstances or situations that are included or excluded
- A list of the incidents identified, together with their prioritization
- The organizational structure for the DR team, with a clear definition of roles, responsibilities, and levels of authority
- Communications procedures and contact information
- Locations of emergency equipment and supplies
- Locations of spare hardware and tools
- Details of recovery objectives for essential functions and resources covered by the DR plan

Recovery Objectives

The DR plan must identify the recovery objectives for each essential function and resource that has been identified in the BIA. There are two key recovery objectives to identify:

1. **The recovery time objective (RTO)** – The duration of time and a service level within which a process must be restored after a disaster (or disruption) in order to avoid unacceptable consequences associated with a break in service

2. **The recovery point objective (RPO)** – The acceptable amount of data loss measured in time (e.g., data must be restored from within 2 hours of a disaster for the loss of that data to be acceptable)

Once these are identified they can be documented in a table similar to that shown in Table 8-2.

Effective disaster recovery requires the identification of realistic and manageable recovery objectives, supported by the resources required to achieve these objectives. Vague recovery objectives ("as soon as possible") are not helpful in disaster recovery planning. Specifying a recovery objective without supporting that with the necessary resources is equally unhelpful. For example, if Process A/Subsystem 1 has a recovery time objective of 8 hours but spare hardware cannot be obtained until the next day, then this objective is unattainable. Either the objective is too aggressive or spares must be held on-site or nearby.

Table 8-2. Example Recovery and Time Objectives per Subsystem

Process	Subsystem	RTO	RPO
Process A	Subsystem 1	8 h	72 h
	Subsystem 2	36 h	72 h
	Subsystem 3	36 h	72 h
Process B	Subsystem 1	2 h	8 h
	Subsystem 2	4 h	8 h
	Subsystem 3	8 h	8 h
	Subsystem 4	8 h	8 h

Disaster Recovery Drills

The DR plan must be exercised on a periodic basis to verify that:

- The recovery procedures work and the documentation provides sufficient detail.
- The DR team members are familiar with the procedures and understand how to work together.
- The preparatory tasks, such as taking backups, are being performed, and the resources and tools are located as expected.

In order to exercise the plan, varying scenarios should be chosen and the drill should be as realistic as possible. The frequency of DR drills will depend on the organization; however, many critical infrastructure organizations undertake them once or twice a year.

Root Cause Analysis and After Action Reviews

Disaster recovery comes to a close when the following aspects are complete:

- An analysis has been performed to identify the root cause of the incident.
- A report is produced that examines the events from first detection to final recovery.

Root cause analysis (RCA; sometimes called after action review or after action report—AAR) is a method for identifying the underlying cause of an incident. The root cause is the original factor that resulted in the eventual incident, which may require some detailed analysis to determine. If the root cause is

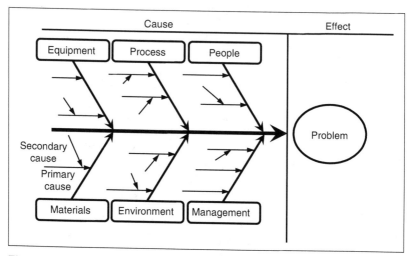

Figure 8-2. Fishbone Diagram for Root Cause Analysis

not identified, the same incident can reoccur. For example, if a fuse blows, causing a loss of power to equipment, and the fuse is replaced without any understanding as to why it blew, then it is likely to reoccur. If an RCA is performed and it is determined that there is a wiring fault, then this must be rectified before replacing the fuse.

There are many methods to document an RCA. One common method is the "fishbone" diagram, as seen in Figure 8-2. In this diagram, the key aspects that may be involved—such as equipment, processes, and people—are listed, and then for each aspect, causes (which can have multiple levels) are listed.

All key players review their notes and verify the IR documentation is accurate and precise. All team members review their actions during the incident, and identify areas where the IR plan worked, did not work, or should be improved. This allows the team to update the IR plan. The RCA can serve as a training case for future staff and brings to a close the actions of the IR team.

Emergency Management

Phases of Emergency Management

The U.S. government identifies four distinct phases in emergency management:

1. **Mitigation** – This includes activities that reduce the likelihood of an emergency occurring or reduce the impact of the effects of the emergency if it does occur. Mitigations could include purchasing insurance or implementing certain cybersecurity controls.

2. **Preparedness** – This includes the plans and preparations that must be performed before an emergency occurs. Maintaining offsite system backups is an example of a preparedness activity.

3. **Response** – This includes the actions that are performed in the event of an emergency. A typical response action is restoring a system from offsite backups.

4. **Recovery** – This includes the activities that are performed after the immediate danger of the emergency is over. This may include replacement of nonessential items that were damaged in the incident.

The U.S. emergency management methodology is designed to engage and coordinate all the available resources at all levels of government and industry. Emergency management plans must take into account that emergencies at mission critical facilities can have serious consequences to the local environment and population. As a result, these plans will need to identify the roles and responsibilities for emergencies, considering:

- Local elected and appointed officials (e.g., governor, mayor, and commissioner)

- Local departments and agencies (e.g., fire, law enforcement, emergency medical services, public health, emergency management, social services, and animal control)

- State agencies (e.g., Department of Transportation, state police/highway patrol, Department of Agriculture, Department of Natural Resources, environmental protection/quality, emergency management, Department of Homeland Security, Department of Health/Public Health, and National Guard)

- Regional organizations or groups

- Federal agencies most often and/or likely to be used to support local operations (e.g., Federal Emergency Management Agency—FEMA, U.S. Coast Guard, U.S. Department of Justice, Federal Bureau of Investigation, Federal Aviation Administration, National Transportation Safety Board, U.S. Department of Defense, U.S. Department of Transportation, and U.S. Department of Agriculture)

- Government-sponsored volunteer resources (e.g., Fire Corps and/or Medical Reserve Corps, Volunteers in Police Service, and Auxiliary Police)

- Private sector and voluntary organizations

Activities that will likely be required in emergency response plans include:

- Population warning and emergency public information
- Public protection
- Mass care/emergency assistance

Review Questions

8.1 What is the name of the plan that identifies the essential functions and resources of an organization?

   A. Business impact analysis (BIA)

   B. Incident response (IR)

   C. Disaster recovery (DR)

   D. Business continuity (BC)

8.2 Which plan identifies how critical functions will continue in the event of a serious incident that requires long-term resolution?

   A. Incident response (IR)

   B. Business impact analysis (BIA)

   C. Business continuity (BC)

   D. Disaster recovery (DR)

8.3 What government body operates the National Incident Management System (NIMS)?

   A. Environmental Protection Agency (EPA)

   B. Department of Homeland Security (DHS)

   C. Nuclear Regulatory Commission (NRC)

   D. North American Electric Reliability Corporation (NERC)

8.4 What is the name of the recovery objective that defines an acceptable amount of data loss in the event of a disaster?

   A. Recovery time objective (RTO)

   B. Recovery point objective (RPO)

C. Response time objective (RTO)

D. Response point objective (RPO)

8.5 What type of diagram is commonly used in a root cause analysis (RCA) process?

A. Incident diagram

B. Bowtie diagram

C. Fishbone diagram

D. Barrier diagram

Appendix A: Answers to Review Questions

Chapter 2: Mission Critical Operations Concepts

2.1 What does the term *availability* mean in relation to systems?

Answer C. The system or information is there when it is needed

2.2 Which of the three aspects are most important for IT and OT systems?

Answer A. For IT systems, it is confidentiality; for OT systems, it is availability

2.3 What is one reason for the convergence of IT and OT systems?

Answer D. The need for greater visibility of plant operation

2.4 What does the term *redundancy* mean in relation to systems?

Answer B. The system continues to operate in the event of a component failure

2.5 What does the term *reliability* mean in relation to systems?

Answer C. The system operates continuously without interruption

Chapter 3: Mission Critical Standards

3.1 What is the key difference between a standard and a regulation?

Answer C. Compliance with regulations is mandatory but compliance with standards is optional

3.2 Which standards body is responsible for IEC 62443, *Security for Industrial Automation and Control Systems*?

Answer B. International Society of Automation (ISA)

3.3 Which standards body is responsible for the Simple Network Management Protocol (SNMP) standard?

Answer D. Internet Engineering Task Force (IETF)

3.4 Which regulatory authority is responsible for the Critical Infrastructure Protection (CIP) regulations?

Answer B. North American Electric Reliability Corporation (NERC)

3.5 Which sector-specific agency is responsible for the Water and Wastewater Systems Sector?

Answer B. Environmental Protection Agency (EPA)

Chapter 4: Mission Critical Technology

4.1 How many layers does the Open Systems Interconnection Reference Model contain?

Answer D. 7

4.2 Which IP addressing standard uses 32 bits to represent a network address?

Answer B. IPv4

4.3 What is the term used to define a network that usually spans several buildings in the same city or town?

Answer B. Metropolitan area network (MAN)

4.4 Why should Wired Equivalent Privacy (WEP) not be used to secure a wireless network?

Answer C. It uses a short encryption key that can be quickly cracked

4.5 What is the name of the network protocol that provides a secure emulation of a traditional computer terminal?

Answer A. Secure Shell (SSH)

4.6 What is the name of the network protocol that is responsible for converting IP addresses to domain names?

Answer D. Domain name system (DNS)

4.7 In a process control network, what is the name of the device that measures a physical value and converts it to a value suitable for processing by a control system?

Answer B. Sensor

4.8 What device is used in a process control network for local monitoring and control?

Answer B. Operator interface terminal (OIT)

4.9 What does a trend display show in a distributed control system (DCS)?

Answer A. Time-series data, such as flow, temperature, pressure, etc.

4.10 What is a key difference between a programmable logic controller (PLC) and a remote terminal unit (RTU)?

Answer A. A PLC is generally used to communicate over local networks, whereas an RTU is used to communicate over wide area networks

4.11 Which of the following process control network communications protocols would most likely be used to communicate with an intelligent electronic device (IED)?

Answer B. DNP3

4.12 What is the name of the process control network communications protocol that is used to configure or calibrate an analog sensor using the same connection that provides the analog measurement?

Answer C. HART, the Highway Addressable Remote Transducer protocol

4.13 What recent innovation allows a single physical server to support multiple parallel operating environments?

Answer D. Virtualization

4.14 Which cybersecurity defense-in-depth protection involves management and control of system and device passwords?

Answer B. Access control

4.15 What are the core activities defined in the NIST Cybersecurity Framework?

Answer D. Identify, protect, detect, respond, and recover

4.16 What type of redundant design involves running duplicate components alongside the live equipment and swapping them in manually in the event of a failure?

Answer B. Warm standby

4.17 What type of encryption scheme involves two keys: public and private?

Answer B. Asymmetric encryption

4.18 What is a good security practice to employ when providing remote access to mission critical systems?

Answer A. Ensure remote access is limited to authorized and competent persons only

4.19 How frequently should mission critical equipment be backed up?

Answer C. Equipment should be backed up so that it can be restored without losing significant data or changes

4.20 What is one key reason why care must be taken when deploying an intrusion detection system (IDS) in a mission critical network?

Answer C. IDSs generate a significant amount of network data that can affect the operation of other networked devices

Chapter 5: Operations

5.1 What is one of the primary purposes for standard operating procedures (SOPs)?

Answer A. To minimize the opportunity for error when undertaking an activity

5.2 What is one good practice that should be employed to maintain good version control of SOPs?

Answer C. Provide a single source of SOP version information to ensure that users know which version of an SOP is to be used

5.3 What is the common term for a good practice to reduce the risk of fraud or human error?

Answer B. Separation of duties

5.4 What does an availability of "five nines" correspond to in terms of downtime in a single year?

Answer D. 5 minutes

5.5 What is the name of the measure that combines quality, performance, and availability to give a single overall score?

Answer B. Overall equipment effectiveness (OEE)

5.6 What is the name of the protocol that is most often used to collect network management data?

Answer A. Simple Network Management Protocol (SNMP)

5.7 What is the key difference between an event and an alarm?

Answer A. Unlike an alarm, an event is something that has occurred that requires no immediate action

5.8 In the context of alarm management, what is a false positive?

Answer B. An alarm was received but there is no incident occurring

5.9 In the RACI matrix, which defines stakeholders in a change management process, what does the abbreviation RACI mean?

Answer C. Responsible, Accountable, Consulted, and Informed

5.10 What is the name of the curve that shows typical equipment failure rates over an entire life cycle?

Answer A. Bathtub curve

Chapter 6: Safety and Physical Security

6.1 What is the collective term for equipment worn by personnel to protect against workplace hazards?

Answer A. Personal protective equipment (PPE)

6.2 On a lockout/tagout (LOTO) tag, what color is used to signify biological hazards?

Answer D. Fluorescent orange

6.3 What is the term for the process used to identify potential hazards in an activity?

Answer C. Job safety analysis (JSA)

6.4 What is the name of the standard that defines the details that must be provided on Safety Data Sheets (SDSs)?

Answer B. Hazard Communication Standard (HCS)

6.5 On a Hazardous Materials Identification System (HMIS) label, what color indicates a physical hazard?

Answer C. Orange

6.6 In process safety, what is the name for the safety feature that makes the state of two mechanisms or functions mutually dependent?

Answer A. Interlock

6.7 In process safety, what is the name of the equipment that is used to manage a particular hazard?

Answer D. Safety instrumented function (SIF)

6.8 In IEC 61508, what is the required availability of a safety integrity level (SIL) 4 SIF?

Answer D. >99.99%

6.9 In U.S. hazardous area classification, what group covers metallic dusts?

Answer C. Group E

6.10 For hazardous area equipment, what is the method that ensures that electrical equipment is incapable of releasing sufficient energy to cause an ignition under normal or abnormal conditions?

Answer A. Intrinsically safe

Chapter 7: Fundamentals of Risk Management

7.1 What are the steps in the risk management life cycle?

Answer A. Identify, assess, control, and review

7.2 What are the key elements of the risk equation?

Answer D. Likelihood and consequence

7.3 What is the activity called that analyzes the potential issues in a system?

Answer C. Vulnerability assessment

7.4 What is a form of risk response that involves purchasing insurance?

Answer D. Transfer

7.5 What is the term that defines the point at which it is not possible to realistically reduce risk any further?

Answer A. As low as reasonably practicable (ALARP)

Chapter 8: Continuity of Operations and Emergency Response

8.1 What is the name of the plan that identifies the essential functions and resources of an organization?

Answer A. Business impact analysis (BIA)

8.2 Which plan identifies how critical functions will continue in the event of a serious incident that requires long-term resolution?

Answer C. Business continuity (BC)

8.3 What government body operates the National Incident Management System (NIMS)?

Answer B. Department of Homeland Security (DHS)

8.4 What is the name of the recovery objective that defines an acceptable amount of data loss in the event of a disaster?

Answer B. Recovery point objective (RPO)

8.5 What type of diagram is commonly used in a root cause analysis (RCA) process?

Answer C. Fishbone diagram

Bibliography

Technology

Battikha, N. E. *The Condensed Handbook of Measurement and Control.* 4th ed. Research Triangle Park, NC: ISA (International Society of Automation), 2018.

Cole, Eric. *Network Security Bible.* 2nd ed. Indianapolis, IN: Wiley Publishing, 2009.

Comer, Douglas. *Computer Networks and Internets.* 6th ed. London: Pearson, 2014.

ISA (International Society of Automation). *The Automation, Systems, and Instrumentation Dictionary.* Research Triangle Park, NC: ISA, 2003.

Krutz, Ronald L. *Industrial Automation and Control System Security Principles: Protecting the Critical Infrastructure.* 2nd ed. Research Triangle Park, NC: ISA (International Society of Automation), 2017.

Langer, Ralph. *Robust Control System Networks*. New York: Momentum Press, 2012.

McAulay, Tyson, and Bryan Singer. *Cybersecurity for Industrial Control Systems*. Boca Raton, FL: CRC Press, Taylor & Francis Group, 2012.

Rhodes-Ousley, Mark. *Information Security: The Complete Reference*. 2nd ed. Columbus, OH: McGraw Hill Education, 2013.

Shaw, William T. *Cybersecurity for SCADA Systems*. Tulsa, OK: PennWell Corporation, 2006.

Thompson, Lawrence (Larry) M., and Tim Shaw. *Industrial Data Communications*. 5th ed. Research Triangle Park, NC: ISA (International Society of Automation), 2016.

Trevathan, Vernon L. *A Guide to the Automation Body of Knowledge*. Research Triangle Park, NC: ISA (International Society of Automation), 2006.

Operations

Cable, Mike. *Calibration: A Technician's Guide*. Research Triangle Park, NC: ISA (International Society of Automation), 2005.

Curtis, Peter M. *Maintaining Mission Critical Systems in a 24x7 Environment*. 2nd ed. IEEE (Institute of Electrical and Electronics Engineers). Hoboken, NJ: John Wiley and Sons, 2011.

Gifford, Charlie, ed. *The Hitchhiker's Guide to Operations Management: ISA-95 Best Practices Book 1.0*. Research Triangle Park, NC: ISA (International Society of Automation), 2007.

Goettsche, Lawrence D. *Maintenance of Instruments & Systems.* 2nd ed. Research Triangle Park, NC: ISA (International Society of Automation), 2005.

Gustin, Joseph F. *Disaster & Recovery Planning: A Guide for Facility Managers.* 6th. ed. Lilburn, GA: The Fairmont Press, 2013.

Hollifield, Bill R., and Eddie Habibi. *Alarm Management: A Comprehensive Guide.* 2nd ed. Research Triangle Park, NC: ISA (International Society of Automation), 2011.

Mostia Jr., William L. *Troubleshooting: A Technician's Guide.* Research Triangle Park, NC: ISA (International Society of Automation), 2006.

Patton, J. D., Jr. *Preventative Maintenance.* 3rd ed. Research Triangle Park, NC: ISA (International Society of Automation), 2004.

Tucker, Eugene. *Business Continuity from Preparedness to Recovery: A Standards Based Approach.* Waltham, MA: Butterworth-Heinemann, 2015.

Vergon, Terry R. *Building Mission Critical Facilities Organizations When Failure Is Not an Option.* lulu.com, 2011.

Safety and Physical Security

Earley, Mark W., ed. *National Electric Code Handbook 2014.* Quincy, MA: National Fire Protection Association (NFPA), 2013.

Goetsch, David L. *The Basics of Occupational Safety.* 2nd. ed. Yorkshire, U.K.: Pearson, 2015.

Griffin, Roger D. *Principles of Hazardous Materials Management.* Boca Raton, FL: CRC Press, Taylor & Francis Group, 2009.

Gruhn, Paul, and Harry L. Cheddie. *Safety Instrumented Systems: Design, Analysis, and Justification*. Research Triangle Park, NC: ISA (International Society of Automation), 2006.

Risk Management

Garvey, Paul R. *Analytical Methods for Risk Management: A Systems Engineering Perspective*. Boca Raton, FL: CRC Press, Taylor & Francis Group, 2009.

Rausand, Marvin. *Risk Assessment: Theory, Methods, and Applications*. Hoboken, NJ: John Wiley & Sons, 2011.

Emergency Response

Erickson, Paul A. *Emergency Response Planning: For Corporate and Municipal Managers*. 2nd ed. Oxford, U.K.: Elsevier Butterworth-Heinemann, 2006.

Gifford, Charlie. *The MOM Chronicles: ISA-95 Best Practices Book 3.0*. Research Triangle Park, NC: ISA (International Society of Automation), 2013.

Haddow, George D., Jane A. Bullock, and Damon P. Coppola. *Introduction to Emergency Management*. 6th ed. Cambridge, MA: Butterworth-Heinemann, 2017.

Lindell, Michael K., Carla S. Prater, and Ronald W. Perry. *Principles of Emergency Management*. Hoboken, NJ: John Wiley & Sons, 2015.

Index

3DES (Triple DES), 66

Access layer, 32
Actuator Sensor Interface (AS-i), 56
Actuators, 25, 44, 107
Advanced Encryption Standard (AES), 66
Address Resolution Protocol (ARP), 37
After action reviews, 171
Alarms, 92–96
    correlation, 95
    display, 93
    events, 95
    escalation, 93
    handling, 93
    priorities, 92
    states, 94
    true/false negatives, 95
    true/false positives, 95
Alerts, 92
Analytics, 57
Analyzer, 26
As low as reasonably practicable (ALARP), 157
Asset register, 97
Asymmetric key algorithms, 67
ATEX, 132

Authentication, 34, 36
    remote, 36
Availability, 6, 87

Bathtub curve, 106
    Early-failure period, 106
    Userful-life period, 106
    Wearout period, 106
Big data, 57
Building Automation and Control Network (BACnet), 56
Business resiliency, 161
Business systems, integrating with, 26

Calibration, 105
Closed-loop control, 50
Cloud computing, 57
Cold standby, 64
Commercial off-the-shelf (COTS), 27, 56
Commissioning, 105
Compliance, 13
Condition monitoring, 107
Confidentiality, 6
Configuration management (CM), 97
Consequence, 148
Continuity of operations (COO), 161

191

Continuous emission monitoring
systems (CEMs), 136
Control methodologies, 49–51
Control room, 46
Control systems, 1, 44
Controller Area Network (CAN), 56
Controllers, 25, 45–49, 55, 56, 65
Core layer, 32
Critical infrastructure, 8
interdependency, 8–10
Cyber attack, 2
Cyber hygiene, 69
Cybersecurity, 58–73
ANSI/ISA-62443, 63
barrier model, 59, 61
cyber attack, 2
cyber hygiene, 69
defense in depth, 59–61
Dover Castle, 59, 60
encryption, 64–68
Executive Order 13636, 62
intrusion detection system
(IDS), 72
management systems, 61
NIST Cybersecurity Framework,
62, 63
redundancy, 63
remote access, 68
Swiss cheese model, 59, 61
USB drives, 71

Data Encryption Standard (DES), 66
Data logger, 25
Defense in depth, 59–61
Diffie-Hellman, 67
Digital Signature Standard (DSS), 67
Disaster recovery drills, 171
Discharge monitoring report
(DMR), 137
Distributed control system (DCS), 45
Distribution layer, 32
DNP3, 54
Domain name system (DNS), 39
Domain registrars, 39
Dynamic Host Configuration Protocol
(DHCP), 38

Early-failure period, 106
Emergency response, 161
after action reviews, 171
business continuity planning, 162
business impact analysis, 162
business resilience planning, 161
disaster recovery drills, 171
disaster recovery, 163
emergency management, 173
emergency scenarios, 109
incident command system, 166
incident response, 163, 173
National Incident Management System
(NIMS), 166
recovery point objective (RPO), 170
recovery time objective (RTO), 169
recovery, 169, 173
root cause analysis, 171
training, 163
Emergency scenarios, 109
Emergency Shutdown System (ESD), 2, 52
Encryption, 34, 64–68
Advanced Encryption Standard
(AES), 66
asymmetric, 66
ciphertext, 65
cleartext, 64
Data Encryption Standard (DES), 66
decrypt, 64
Diffie-Hellman, 67
Digital Signature Standard (DSS), 67
encryption key, 35, 65
Internet Key Exchange, 68
IPsec, 68
plaintext, 64
private key, 66
public key, 66
RSA, 67
symmetric, 65
Triple DES (3DES), 66
Enhanced Interior Gateway Routing
Protocol (EIGRP), 40
EtherNet/IP, 54
Events, 92, 95, 121
Extensible Authentication Protocols
(EAP), 36

Fail-safe mechanisms, 127, 128
Failure modes and effects analysis (FMEA), 129
Fault tree analysis (FTA), 129
Fishbone diagram, 172
"Five nines," 87
Forecasting, 104
FOUNDATION Fieldbus, 55

Hazard and operability study (HAZOP), 127
Hazard Communication Standard (HCS), 121, 123
Hazard, 148
Hazardous area zone, 131, 134
   equipment protection methods, 136
   temperature, 135
Hazardous Materials Identification System (HMIS), 124
Hazardous process controls, 126
Hazards, to workplace personnel, 115, 120
Heating, Ventilation, and Air Conditioning (HVAC), 9
Highway Addressable Remote Transducer (HART), 55
Historian, 52
HMIS sign, 124
Host-based intrusion detection system (HIDS), 72
Hot standby, 64
Human-machine interfaces (HMIs), 46

IEC 61508, 130
IEEE 802.1X, 36
Incident command system, 166
Industrial Internet of Things (IIoT), 57
Information technology (IT), 2, 5–10
Internet Control Message Protocol (ICMP), 39
Interior Gateway Protocol (IGP), 40
Inspection, 105
Integrity, 6
Integrated Control and Safety System, 52
Intelligent electronic device (IED), 54
Interlock, 128

International Society of Automation (ISA), 16, 63, 131
International Telecommunications Union (ITU-T), 27
International Telegraph and Telephone Consultative Committee (CCITT), 27
Internet Control Message Protocol (ICMP), 39
Internet Corporation for Assigned Names and Numbers (ICANN), 39
Intrusion detection system (IDS), 72, 73
Intrusion prevention system (IPS), 73
IP addressing, 29–31
   classes of, 30
   classless interdomain routing (CIDR), 30
   dynamic, 38
   IPv4, 29, 31
   IPv6, 29, 31
   nonroutable addresses, 30
   static, 38
IPsec, 68
ISA84, 131

Job hazard analysis, 120
Job safety analysis (JSA), 120

Kanban, 105
Key performance indicator (KPI), 89

Layer of protection analysis (LOPA), 129
Life-cycle management, 104
   commissioning, 105
   condition monitoring, 107
   decommissioning, 108
   forecasting and provisioning, 104
   just-in-time manufacturing, 105
   lean manufacturing, 105
   plant shutdown, 108
Likelihood, 148
Linux, 57
Local area network (LAN), 33
Lockout/Tagout (LOTO), 118
   procedures, 119
   tags, 119
Loss of containment, 137

Maintainability, 88
Maintenance turnaround, 108
Management information base (MIB), 92
Management of change (MOC), 100
Material safety data sheets (MSDSs), 121
Media access control (MAC), 37
Method statement, 100
Metropolitan area network (MAN), 33
Mission critical, 1
Mitigation, 148, 173
Modbus, 44, 53

National Electrical Code (NEC), 132
National Incident Management System (NIMS), 166
National Institute of Standards and Technology (NIST), 17, 62, 63
Network management system (NMS), 92
Network-based intrusion detection system (NIDS), 72

Obsolescence, 101, 108
Occupational Safety and Health Administration (OSHA), 18, 119, 121, 129
Occupational safety, 115
Open-loop control, 50
Open Shortest Path First (OSPF), 40
Open source, 57
Open systems interconnection (OSI) model, 27
Operational technology (OT), 2, 5–10
Operations, 2, 5, 81
    alarms, 92–96
    audits, 99
    critical repairs, 90
    downtime, 87
    "five nines," 87
    elements of, 84
    key performance indicator (KPI), 89
    maintenance window, 88
    management of change (MOC), 100
    method statement, 100
    overall equipment effectiveness (OEE), 89
    performance objectives, 87, 89
    permit to work (PTW), 102
    recovery point objective (RPO), 89
    recovery time objective (RTO), 89
    redline drawings, 103
    service level agreement (SLA), 89
    standard operating procedure (SOP), 82–86
        version control, 82
    toolbox talks, 103
    troubleshooting, 88
Operator interface terminals (OITs), 47
Operator, 48
Overall equipment effectiveness (OEE), 89

Permit to work (PTW), 102
Personal protective equipment (PPE), 115–117
Plant outage, 108
Plant shutdown, 108
Plant turnaround, 108
Port number, 41
Preparedness, 173
Presidential Policy Directive, 21, 21
Pressure relief valve (PRV), 128
Process conditions, 25
Process control networks, 44–56
    actuators, 44
    alarm list, 46
    distributed control system (DCS), 45
    flow meter, 45
    Operator interface terminals (OITs), 47
    plant overview, 45
    Proportional-integral-derivative (PID) algorithm, 51
    sensors, 44
    supervisory control and data acquisition (SCADA), 47
    trend display, 45
Process hazard analysis (PHA), 127, 152
Process safety management (PSM), 129
Process safety, 126
PROFIBUS, 54
PROFINET, 54
Programmable logic controller (PLC), 25
Proportional-integral-derivative (PID) algorithm, 51

Protocols, 37–44
  Address Resolution Protocol (ARP), 37
  application layer, 37
  Dynamic Host Configuration Protocol (DHCP), 38
  Enhanced Interior Gateway Routing Protocol (EIGRP), 40
  Exterior Gateway Protocol (EGP), 40
  File Transfer Protocol (FTP), 42
  Hypertext Transfer Protocol (HTTP), 43, 94
  Interior Gateway Protocol (IGP), 40
  Internet Control Message Protocol (ICMP), 39
  networking, 37, 39
  Open Shortest Path First (OSPF), 40
  redundancy, 37, 41
  Reverse Address Resolution Protocol (RARP), 37
  Routing Information Protocol, 40
  routing, 37
  Secure Shell (SSH), 42
  Simple Mail Transfer Protocol (SMTP), 43
  Simple Network Management Protocol (SMNP), 92
  Telnet, 43
Provisioning, 104

Qualitative risk analysis, 152
Quantitative risk analysis, 152

RACI, 101, 102
RASCI, 101
Recovery point objective (RPO), 170
Recovery time objective (RTO), 169
Redundancy, 10
Regulations, 13, 18–23
  common regulatory bodies, 18
  compliance, 82
  Department of Health and Human Services (HHS), 21
  Department of Homeland Security (DHS), 21
  Environmental Protection Agency (EPA), 19
  Federal Energy Regulatory Commission (FERC), 19
  Food and Drug Administration (FDA), 20
  North American Electric Reliability Corporation (NERC), 19
  Nuclear Regulatory Commission (NRC), 19
  Occupational Safety and Health Administration (OSHA), 18, 119, 121, 129
  Payment Card Industry (PCI) Security Standards Council, 21
  Pipeline and Hazardous Materials Safety Administration (PHMSA), 20
  Securities and Exchange Commission (SEC), 21
Relational data, 52
Reliability, 10, 87
Remote access, 68
Remote Authentication Dial-In User Service (RADIUS), 36
Resilience, 10
Resource leveling, 109
Reverse Address Resolution Protocol (RARP), 37
Risk reduction factor (RRF), 130
Risk, 145
  ALARP, 157
  analysis, objective of, 156
  analysis, qualitative, 152
  analysis, quantitative, 152
  consequence scale, 150
  key components of, 148
  process hazard analysis (PHA), 127, 152
  response, 149
  risk management cycle, 146
  risk matrix, 151
  Risk reduction factor (RRF), 130
  safety definition, 145
  security definition, 146
  threat assessment, 154
  tolerance, 149
  vulnerability assessment, 148, 155
Root cause analysis, 171

Routing Information Protocol, 40
RS-232, 54
RS-485, 54
RSA (Rivest, Shamir and Adleman), 67
Rupture disk, 128

Safety data sheets (SDSs), 121
Safety instrumented function (SIF), 130
Safety Instrumented System (SIS), 52
Safety Shutdown System (SSDS), 52
Safety orientation, 119
Safety, occupational, 115
   Hazard Communication Standard (HCS), 121, 123
   Hazardous Materials Identification System (HMIS), 124
   job hazard analysis (JHA), 120
   job safety analysis (JSA), 120
   Lockout/Tagout (LOTO), 118
   material safety data sheets (MSDSs), 121
   orientation, 119
   Personal protective equipment (PPE), 115–117
   safety data sheets (SDSs), 121
Safety, process, 126
   equipment protection methods, 136
   fail-safe mechanisms, 127, 128
   failure modes and effects analysis (FMEA), 129
   fault tree analysis (FTA), 129
   hazard and operability study (HAZOP), 127
   hazardous area zone, 131, 134, 136
   hazardous process controls, 126
   layer of protection analysis (LOPA), 129
   loss of containment, 137
   process hazard analysis (PHA), 127, 152
   safety instrumented function (SIF), 130
Secure Shell (SSH), 42
Sensors, 25, 44, 107
Service continuity, 161
Service level agreement (SLA), 89
Session key, 68
Simple Mail Transfer Protocol (SMTP), 43
Simple Network Management Protocol (SMNP), 92
Spare parts, 109
Standard operating procedure (SOP), 82–86
Standards, 13–23
   American National Standards Institute (ANSI), 13
   American Petroleum Institute (API), 15
   American Society of Heating, Refrigerating and Air Conditioning Engineers (ASHRAE), 14
   American Society of Mechanical Engineers (ASME), 15
   common standards bodies, 14
   compliance, 13
   Institute of Electrical and Electronics Engineers (IEEE), 16
   Interagency Security Committee (ISC), 16
   International Electrotechnical Commission (IEC), 15
   International Organization for Standardization (ISO), 17, 27
   International Society of Automation (ISA), 16, 131
   Internet Engineering Task Force (IETF), 16
   National Electrical Manufacturers Association (NEMA), 17
   National Fire Protection Association (NFPA), 17
   National Institute of Standards and Technology (NIST), 17, 62
   Standard operating procedures (SOPs), 83
   Standards development organization (SDO), 13
   U.S. Code of Federal Regulations (CFR), 18
Static IP address, 38

Supervisory control and data acquisition (SCADA), 47
   alarm display, 93–95
Symmetric key algorithms, 66
System hardening, 70

Telnet, 43
Testing, commissioning, 105
Threat assessment, 148, 154, 155
Time-series data, 52
Transmission control protocol/internet protocol (TCP/IP) model, 28
Triple DES (3DES), 66

Troubleshooting, 88

Useful-life period, 106, 108

Virtual private network (VPN), 34
   gateway, 35
Virtualization, 57
Vulnerability assessment, 148, 155

Warm standby, 64
Wearout period, 106
Wide area network (WAN), 34
Windows Management Instrumentation (WMI), 91